CAMBRIDGE LIBRARY COLLECTION

Books of enduring scholarly value

Zoology

Until the nineteenth century, the investigation of natural phenomena, plants and animals was considered either the preserve of elite scholars or a pastime for the leisured upper classes. As increasing academic rigour and systematisation was brought to the study of 'natural history', its subdisciplines were adopted into university curricula, and learned societies (such as the London Zoological Society, founded in 1826) were established to support research in these areas. These developments are reflected in the books reissued in this series, which describe the anatomy and characteristics of animals ranging from invertebrates to polar bears, fish to birds, in habitats from Arctic North America to the tropical forests of Malaysia. By the middle of the nineteenth century, this work and developments in research on fossils had resulted in the formulation of the theory of evolution.

The Life and Letters of Gilbert White of Selborne

Published in 1901, this illustrated two-volume biography of the renowned English naturalist Gilbert White (1720–93) presents a thorough account of his life and achievements. Prepared by White's great-great-nephew Rashleigh Holt-White (1826–1920), it incorporates a selection of White's correspondence with family and friends, providing valuable insights into his beliefs and character. Included are letters sent by White's lifelong friend John Mulso (1721–91), who praised the naturalist's work, predicting it would 'immortalise' White and his Hampshire village. Still considered a classic text, *The Natural History and Antiquities of Selborne* (1789), featuring White's careful observations of local flora and fauna, is also reissued in the Cambridge Library Collection. In the present work, Holt-White sought to correct the 'erroneous statements' that had previously been made about his relative. Volume 2 traces White's life from September 1776, considering the impact of the loss of family members, and his legacy after his death.

Cambridge University Press has long been a pioneer in the reissuing of out-of-print titles from its own backlist, producing digital reprints of books that are still sought after by scholars and students but could not be reprinted economically using traditional technology. The Cambridge Library Collection extends this activity to a wider range of books which are still of importance to researchers and professionals, either for the source material they contain, or as landmarks in the history of their academic discipline.

Drawing from the world-renowned collections in the Cambridge University Library and other partner libraries, and guided by the advice of experts in each subject area, Cambridge University Press is using state-of-the-art scanning machines in its own Printing House to capture the content of each book selected for inclusion. The files are processed to give a consistently clear, crisp image, and the books finished to the high quality standard for which the Press is recognised around the world. The latest print-on-demand technology ensures that the books will remain available indefinitely, and that orders for single or multiple copies can quickly be supplied.

The Cambridge Library Collection brings back to life books of enduring scholarly value (including out-of-copyright works originally issued by other publishers) across a wide range of disciplines in the humanities and social sciences and in science and technology.

The Life and Letters of Gilbert White of Selborne

VOLUME 2

EDITED BY
RASHLEIGH HOLT-WHITE

CAMBRIDGE
UNIVERSITY PRESS

CAMBRIDGE
UNIVERSITY PRESS

University Printing House, Cambridge, CB2 8BS, United Kingdom

Cambridge University Press is part of the University of Cambridge.
It furthers the University's mission by disseminating knowledge in the pursuit of
education, learning and research at the highest international levels of excellence.

www.cambridge.org
Information on this title: www.cambridge.org/9781108076494

This edition first published 1901
This digitally printed version 2015

ISBN 978-1-108-07649-4 Paperback

THE LIFE AND LETTERS OF
GILBERT WHITE OF SELBORNE

Bernard Lens, pinx.

John White

Walker & Cockerell, ph.sc.

THE LIFE AND LETTERS OF
GILBERT WHITE
OF SELBORNE

WRITTEN AND EDITED BY

HIS GREAT-GRAND-NEPHEW

RASHLEIGH HOLT-WHITE

WITH PEDIGREE, PORTRAITS, AND ILLUSTRATIONS

IN TWO VOLUMES

VOL. II.

LONDON

JOHN MURRAY, ALBEMARLE STREET

1901

CONTENTS

CHAPTER IV.

CHAPTER V.

CHAPTER VI.

CHAPTER VII.

CHAPTER VIII.

LIST OF ILLUSTRATIONS

THE LIFE AND LETTERS OF
GILBERT WHITE
OF SELBORNE

CHAPTER I.

On September 27th, 1776, Mulso writes that he was unable to visit his friend this year, because—

"I have a great variety of businesses now on my hands, . . . I have all my farmers to compose. . . . I am horribly provoked, for curiosity as well as affection draw me towards you: and Mrs. Mulso, hearing of practicable egress and regress, remits her apprehensions of walking to Selborne: as to my girls, the thought of it is a banquet. . . . Oh! how unlike is the visit of Bloxam * and attorney Knott to the elegant attendance of Mr. Grimm, who came to perpetuate scenery so dear to you? Yours is a life of virtù, and mine of carking and caring."

The influence of Mr. Grimm is perhaps apparent in the following entry in the *Naturalist's Journal* :—

"Oct. 12. The hanging beech-woods begin to be beautifully tinged, and to afford most lovely scapes very engaging to the eye and imagination. They afford sweet lights and

---

* A surveyor.

shades. Maples are also finely tinged. These scenes are
worthy of the pencil of a Rubens."

"Nov. 1. [Fyfield.] Four swallows were seen skimming
about in a lane below Newton. This circumstance seems
much in favor of hiding, since the *Hirundines* seemed to
be withdrawn for some weeks. It looks as if the soft
weather had called them out of their retirement."

From Fyfield his uncle wrote—

To Samuel Barker.   Fyfield, Novr. 1, 1776.

Dear Sam,—Just as I thought you had been master of the
manners and customs of the bank-martin, you write me
word that you do not know it when you see it. The case is,
you did not begin to look 'til the decline of summer, when
all the *Hirundines* cease to frequent their nesting-places.
If you will pay some attention to those holes in the spring,
you will probably see the owners busyed in the matters of
nidification: besides they are to be distinguished from their
congeners by their *small size*, their *mouse-colour*, and their
wriggling, desultory manner of *flying*. Pray observe when
they come first.

The instance you give of the swiftness of an hawk was
somewhat extraordinary. But a very intelligent person once
assured me that he saw a more extraordinary instance of
command of wing in a daw, which is not very remarkable
for feats of activity of that kind. As this person was riding
on Salisbury plain he saw a bird on the wing dropping
something from its bill, and catching it again before it came
to the ground, several times repeatedly: this unusual sight
drew his attention, so that he rode nearer, and saw still the
same feat repeated to his great surprize. It appeared to him
that the ball dropped and recovered was a wallnut. Now a
wallnut, I should think, would fall much faster than a dead

bird, whose feathers would meet with resistance from the air.

In 24 days Mr. Grimm finished for me 12 drawings; the most elegant of which are 1, a view of the village and hanger from the short Lithe *; 2, a view of the S.E. end of the hanger and its cottages, taken from the upper end of the street; 3, a side view of the *old* hermitage, with the Hermit standing at the door,† this piece he is to copy again for Uncle Harry; 4, a sweet view of the short Lithe and Dorton from the lane beyond Peasecod's house. He took also two views of the Church‡; two views of my outlet; a view of the Temple-farm§; a view of the village from the inside of the present hermitage; Hawkley hanger, which does not prove very engaging; and a grotesque and romantic drawing of the water-fall in the hollow bed of the stream in Silkwood's vale to the N.E. of Berriman's house. You need not wonder that the drawings you saw by Grimm did not please you; for they were 3s. 6d. pieces done for a little ready money: so there was no room for softening his trees, &c. He is a most elegant colourist; and what is more, the use of these fine natural stainings is altogether his own; yet his pieces were so engaging in Indian-ink that it was with regret that I submitted to have some of them coloured. Mr. Wyndham of Sarum has engaged Grimm next summer for eight or nine weeks in a tour round N. and S. Wales. I rejoice to hear you are so deep in French.

I am wonderfully delighted, with the addition to my Brother's little common parlor; "Nequeunt expleri corda tuendo"; it now altogether gives much ease to such a numerous family; and is very peculiar, light, roomy and

* The large folding frontispiece to 'The Natural History and Antiquities of Selborne,' first edition.
† The vignette on the title-page, *op. cit.*
‡ Placed opposite pp. 315, 323, *op. cit.*
§ Placed opposite p. 342, *op. cit.*

convenient, containing 450 square feet. Mr. Amyand is a very genteel pleasing youth; he puts me in mind of Mr. Brocket.   Yours affect.,
                             GIL. WHITE.
Pray write soon to Selborne.

When the children are buzzing down at their spinnet, and we grave folks sit round the chimney, I am put in mind of the following couplet, which you will remember :—

"... all the distant din that world can keep
Rolls o'er my grotto, and improves my sleep."

It is very extraordinary that the new chimney does not smoke in the least. Mr. Henry Woods has been at my house, and has taken his daughter to Chichester; so there are now no children at nurse at Selborne or Newton!

To the Rev. John White.   Fyfield, Novr. 2, 1776.

Dear Brother,—As you have experienced so often how very necessary exercise is for your health, you will no doubt be careful how any avocation or pursuit, how laudable soever, shall again interrupt that regimen so essentially needful. Our brother Thomas has found vast benefit from his journey to Bath: the waters, and the bathing have quite removed for the present both his internal and external ails. He advises, I find, if your rheumatism returns, a journey to Buxton.

Jack is very tall indeed! but if he continues healthy, it will be esteemed an advantage to be a well-grown man. You have never told me whether he was bound for five or seven years.

With respect to your MS. you seem a cup too low; and do not assume the importance of an author. If Mr. Pennant had got such a work ready, he would feel little diffidence; and would expect it would produce some money. If you desire it, I shall be willing to look it over; and

HENRY WHITE'S HOUSE AT FYFIELD

perhaps brother Thomas will do the same when at leisure. By what I saw perhaps some articles may be thought too long. The whale-fishery is a fine new circumstance, and worthy of a national attention; especially as we may soon possibly have nothing to do with the N. American seas. But in such narrow limits, and so warm a climate, how can such an offensive occupation be carryed on without proving a vast nuisance to the garrison? Train-oil, and whales flesh must smell very vigorously in lat. 36. How wise have all the Naturalists proved themselves to be by laying it down for granted that there were no whales in the Mediterranean.

Brother Harry has now a fine annual income; and will, I trust now, when he comes to rest a little from his labours of building, be able to lay up money for his family. His new pupil is a very pleasing, ingenuous youth. Mr. Halliday, though upwards of six feet high, does not leave him 'til Feb. We are now sitting in the new edition of his old parlor enlarged; it is odd and peculiar, but very roomy, light and convenient, and every way suitable to a vast family.

The ceiling of the new part is eleven feet high. This new building is thirty feet long; so that my sister gets the N. end for a store-room: and over all are to be two small lodging-rooms; so there are now in this house twelve chambers. The whole room is wainscotted with deal.*

Last night my brother received a letter from the attorney near Manchester, who wishes to be curate of Darwen. He is urgent for matters to be brought to a bargain. Sure the injunctions and provisions against simony have never reached your part of the world. If disappointed he will not, I hope, stir up a clamour against the southern non resident.

---

* Henry White's house at Fyfield still exists, apparently not much altered externally. It stands near the Rectory (partially rebuilt in 1830), which he used as a schoolhouse. His enlarged parlour cannot now be certainly identified owing to internal alterations.

Dick* is with me; he is good-natured, and some what heady at times. It is well he is intended for trade, since he loves anything better than book: bodily labor he does not spare; for rolling, wheeling, water-drawing, grass-walk-sweeping are his delight. I have taught him to ride; and perhaps a good seat on an horse may be more useful to him than Virgil, or Horace. I tryed Phædrus; but my patience failed. However he may procure health and strength, and a little behaviour at my house.

We all join in respects. My brother's outlet is still pleasing.   Yˣˣ affect.,

GIL. WHITE.

Early in the year 1777 Gilbert White visited his brother Thomas in London, whence he wrote—

To the Rev. John White.
                                    Thames street, Feb. 27 [1777].

Dear Brother,—Many thanks for your letter of the 18th and for your extract from Reaumur. We all much approve of what you intend to *inscribe* to the Archbishop, thinking it neat and polite: but like yourself we do not much like your title-page. Brother Ben. says he thinks that 'Hist. nat. observations in Lat. 36' should all be left out; and that it should begin with 'An Essay,' &c., but it is not worth while to be solicitous about a title-page: Swift says, "for a title-page consult your bookseller." But the term 'Fauna Calpensis,' tho' judged to be too quaint and pedantic for the beginning of a title, yet, I think, must by no means be sunk for the following reason, because I believe you have always told Linnæus that you should call your book by that name; and therefore if he mentions your work in his last edition (as he certainly will) you will lose all the credit

* His nephew Richard, son of Benjamin White, now aged 15.

to be derived from such notice of you, if you mention no such title. Supposing Linn. to be dead, there can be no doubt but that his son will put forth the new edition. By what we remember of the specimen of your work, we thought some articles too diffuse. It is natural for you to fall a little into this extreme from the regard you express for Reaumur; since all the French in Natural History are very circumstantial. Be so good as not to forestall my cobweb-shower;* I wish I had two or three dozen more of such anecdotes. An engraver has been with me; and I have been talking with him about his taking off five or six of my drawings: he says that my quarto drawings cannot be well executed under eight guineas a piece: now five times eight is forty! Grimm is reducing my Hermitage-view in order to bring it to a proper size for a *vignette:* he is also to take it in a large scale for brother Henry. You will see in the papers a remarkable cause in the Commons between a Patron and a Rector who took two *distant* perpetual curacies: the matter was determined in favor of the Rector; had it gone against him the Rector of Fyfield would have had cause to quake. I propose staying in town 'til the 14th of March. Respects to my sister.

Your aff. brother,

GIL. WHITE.

If you think the mention of your degree of A.B. will occasion any inconvenience you may easily drop it. Brother Thomas waits on the Dean of Ely to-morrow at Lambeth: and will be sure to desire him to represent you and Harry in a favourable light to the new Bishop of Chester. Poor Nanny White † declines very fast, and is in a very languishing state.

* *Vide* 'The Natural History and Antiquities of Selborne,' Letter XXIII. to Pennant.
† Daughter of Benjamin White, senr.

A somewhat serious attack of illness occurred during the London visit.

To the Rev. John White.   Selborne, May 2 [1777].

Dear Brother,—I should have wished that you had found your book more marketable, and that you could have sold it *outright*. Yet if Benjn offers to *join*, it looks as if he did not fear the want of success in the publication: besides booksellers have ways and means of subscribing off among the trade in which authors cannot avail themselves.

My thanks are due for your calling on Edm. Woods, who will, I think, soon supply me with some windows.

I wish I could prevail on you to come down and spend a little time with us before you return Northward.

As soon as I got to town I sent your *Hortus siccus* by my brother Thomas's boy to Mr. Curtis's own house; and was in hopes he would have examined the plants.

No swifts appear yet, though we have soft weather.

My left hand is full of gout: all my fingers look red, and shoot and burn. If I have gout about me it is best to come out. I hope you found and left Mrs. Snooke well.

The spirit of building prevails much in this district; Richd Butler, the thatcher, is going to enlarge his house; John Bridger of Oakhanger builds a new one next spring; and Mr. P.[owlett] of Rotherfield began pulling down yesterday.

> "The child that is unborn may rue
> The *pulling* of that day."

I am your loving brother,

GIL. WHITE.

Pray write often, and let me hear what steps you take respecting your book.

On June 1st, 1777, Mulso wrote :—

"It was from Dr. Balguy's information, (who returned to us very lately) that I learnt you had been very dangerously ill in London; but my comfort came at the same time, for he mentioned your being recovered, and attending at the visitation at Alton. . . . I am curious to know whether the regimen that you must have been put into for your cure, had any effect on your deafness. . . . As I do not see any advertisements in the papers, I conclude by the time of year that you have deferred your publication till next winter. I wish you had not: your brother Ben. is a timid man, and you yourself are too modest and nice. The humour for such performances will be over, and make something against the merit of even your book. I felt impatient to see it, with the decorations of Mr. Grimm."

From the *Naturalist's Journal*—

"June 6. Began to build the walls of my parlor, which is 23 ft. and half by 18 ft.; and 12 feet high and 3 inches."

To Thomas White.

[With a copy of a letter from Dr. Chandler* respecting The Temple, near Selborne.]

Selborne, June 20, 1777.

Dear Brother,—The Doctor's letter on the other side is very satisfactory, and very edifying: for it not only proves that *our Temple* belonged to the *Knights Templars*; but that it was also a *preceptory*, the PRECEPTORY of SUDINGTON; now

* Dr. Richard Chandler (1738-1810) was educated at Winchester College, and Queen's College, Oxford. He became Demy of Magdalen 1757, and Probationary Fellow 1770. In 1764 he was commissioned by the Society of Dilettanti to travel for antiquities in Asia Minor and Greece, of which he published a narrative, as well as other works. In 1777 he was presented to the livings of Worldham and West Tisted, in the immediate neighbourhood of Selborne. In 1790 he settled for some time at Selborne Vicarage. As will be seen, he rendered very material help to Gilbert White in the preparation of his account of the antiquities of Selborne, in which parish Magdalen College owned the Priory Estate.

called *Southington* : notwithstanding Bishop Tanner asserts that he never could find more than *two* preceptories in this county, viz. *Godesfield* and *S. Badeisley.* Hence we may be certain that the Bishop did not get access to the papers in Magdalen College archives.

Though the lands of the convent and the Templars abutted on each other, and were intermixed; yet we see that those two societies of Religious lived on the best of terms, in an intercourse of mutual good offices, exchanging lands, and permitting roads to be opened for each other's mutual convenience.

We see also that *Blackmere* and *Bradshot*, names well known to modern ears, were also familiar to the neighbourhood four or five hundred years ago.

I expect Dr. Chandler soon; and regret much (and he assures me he does the same) that the statutes will not permit him to bring with him the archive-papers to Selborne, which contain much knowledge concerning the antiquities of this place; information that has never been pryed into; but has slumbered within the College-walls ever since they were founded.

We have drowning weather, and a dismal black solstice. Such rains make carriage very irksome, and the attendance on building very comfortless, and brick-burning very precarious : but the walls, I trust, will be the stronger; since the mortar is the better blended into the chinks and crevices during so sloppy a season.

Let me hear how you have sold your oak-timber.

Y^{rs} &c.,

Gil. White.

To the Rev. John White.   Selborne, July 16, 1777.

Dear Brother,—Some how or other I had persuaded myself that you were to write first; and having little to say, as we had seen each other so lately, I thought I would stay 'til you gave the challenge before I attacked you with an epistle.

As yet I have not seen your work; but shall peruse it with pleasure as soon as brother Thomas brings it. But he is going at present to bathe on the coast of Dorset for a few weeks.

As I hoped and expected to see you derive some credit and emolument from your labours, I was sorry to hear that the whole pursuit is thrown aside for the present in some degree of disgust and chagrin. One thing I could never understand, and that is, that you say in a former letter, "that having so near a relation a bookseller, should you not agree with him about terms, no other publisher would meddle with your work, because your relation is one of the first editors in the natural history way": now the force of this argument I could never see: for Cadel[1] or any other man would be influenced alone by his own judgment; and if he saw merit in the work, and an interesting subject, would little regard, I should think, another person's sentiments. Unless you have experienced the inconvenience that you thought you foresaw, your suspicions were probably wrong.

The roof of my great parlor is finished; and my walls in a few days will be up to their proper pitch; so that we shall soon proceed to *rearing*. You do well in removing the earth that lies above your floors: I have taken away much for the same reason.

I have not seen the clergy-act, but am assured that it has nothing to do with *residence*: there is nothing compulsive in it: but it enables the clergy to borrow money on their livings, which they may lay out on the repairs of their houses, &c., and so exempt their representatives at their deaths for heavy dilapidations. For the money borrowed the *resident* incumbent is to pay *five* per cent. and some small proportion of the principal off annually; a *non-resident* must pay *ten* per cent. and when the borrower dies the residue remains a debt on the living 'til by degrees it is payed off. Mr. Etty, as far as he knows of the matter yet, for neither has he seen the act, approves much of the plan, and thinks he may avail himself of the matter so as to save himself from heavy demands on his family at his death. The security to the *lender* seems to be safe and good, since the *living*, not any *particular* incumbent, is answerable: but some think few men that have money will care to lend it so as to have the *principal* repaid at the rate perhaps of only 20*s* per ann. We had wet weather all the month of May: but from the 10th of June to the 9th of July it was the strangest summer-solstice I ever saw; nothing but wind, and floods, and clouds, and wintry doings; so that we kept fires in the parlor most part of the time. We have now sweet weather. Respects to my sister.

Y^{rs} &c.,

GIL. WHITE.

A copy of the following letter from Mr. Sewell, Rector of Headley, to Gilbert White, was forwarded by him to his brother Thomas; no doubt in view of the latter's proposed publication upon Hampshire antiquities. To the letter Gilbert White added the postscript below given :—

Headley, Aug. 7, 1777.

Revd Sir,—Out of a large pot of Medals (about 3 years since) which were found in Wulmere pond, I collected a regular series, from Claudius Drusus to Commodus included ; that is, Medals of all the Roman Emperors from A.D. 43 to 194 with those of the two Faustinas, and Crispina, Empress of Commodus: and after Commodus I found no more. Also among the rest I found that of Trajan's famous Stone Bridge over the Danube below Belgrade ; which if it had

been found, when the three bridges at London were first
planned viz. Westminster, London and Blackfriars Bridges,
would then have been of very great value. Vespasian, a
General under Claudius Drusus, about A.D. 47, marched
down with a Roman Army this way, from the parts where
London now is, towards Porchester, S. Hampton, and the
Isle of Wight. It is beautiful, on Headly Heath and
Common, to observe the Entrenchments of the Romans,
and Britains over against each other: the first advancing,
the other retreating. The Romans crossed Headly River at
Stanford, and advanced to the place, where now is Wulmere
pond; and there fixed an abiding Station or City, which
remained for near on 150 years; when they seem to have
been expelled thence by the Britains, or perhaps by an
earthquake or some other cause. Great treasures even now
lye buried in that pond, of Roman Antiquities, of Coins and
medals, of instruments of war, and husbandry, and various
utensils for various uses.

Of the vast quantity of Medals found there, as you
mention about 40 years since, no kind of historical use was
ever made that I ever heard of: when this plain and obvious
historical Truth might easily from thence have been deduced,
the commencement, continuance, or duration of the Roman
Station, or City of Wulmere in Hants. I believe may be
traced from thence vestiges of Roman Roads to Porchester,
Winton, &c.—The Rt. Hon. Mr. Legg got a great quantity
of these coins; and with him they lye dormant; as also do
a great quantity with Whitehead Esq. of Liphook, and with
Mr. Hugonin. And this is the misfortune of most Antiquities and Curiosities, that they frequently fall into
hands that can collect nothing from them; in whose coffers
they are more buried than if they were to lye in the depth
of a mine, or of Wulmere pond. The greatest curiosity
hereabouts is, as I said, the advancement of the Roman
Army to the S.W., over Hindhead, and over Headly upper

Heath and Common. What may be observed of this kind,
by way of Liphook over Hindhead, I have not yet searched
and examined.   I am, Sir, most respectfully,
                    Your obedient servant,
                                    Wm SEWELL.

" Let those, who weekly, from the city's smoke,
  Croud to each neighb'ring hamlet, there to hold
  Their *dusty* sabbath, tip with gold and red
  The milk-white palisades, that Gothic now,
  And now Chinese, now neither, and yet both,
  Checquer their trim domain."

"These seem to be some of the best lines in Mr. Mason's
'English Garden,' which is a work that has disappointed
me much, especially as the subject and the credit of the
author had raised my expectations. By last post I wrote
to you on an interesting subject, and hope for your answer
soon : I shall now, I trust, be able to secure that '*Angulus*'
which the family have been labouring so long to obtain.
Harry and Cane left me this morning; the former expressed
no uneasiness to me concerning the matters you hint at.
Dr. Chandler is reading brother John's papers with great
attention and assiduity ; but says, in order to their becoming
more popular that they should be thrown quite out of their
systematic arrangement."

From Mulso, August 19, 1777 :—

"I know you to be that sort of man, who is long in
determining upon any point, but constant to the plan
established. I have therefore considered you for some time
as a man plunged into mortar. . . . I like the scheme of con-
farreation between your brother and you. He is a man of
sense and vivacity and will teach the *goût* to be of use to
you. I am not at all surprised at your improvement, even
though you had not had the furtherance of your brother,
for you have been nibbling at it a long time ; and to say

truth I did not know but that this expatiating scheme might depend upon another, and that you was preparing to exhibit to us Benedict the married man. I knew such a venture was too delicate to be explained even to an old friend, till it was quite resolved upon; and then like *January* you would have called your council about you. I hope it is better as it is, though I declare I should have spoken with *Placebo* and not with *Justin*.

"Let me, however, know how matters are going with you; and whether, if an opportunity offered of my calling upon you, I should have nothing but a Hod for a Hammock. I feel awkward, if a summer slips by me, and I do not see Selborne."

On August 30th it is noted—

"Finished tiling the new parlor in good condition."

The brother mentioned by Mulso was undoubtedly brother Thomas, to whose eldest son, Thomas Holt-White, the house and property at Selborne were originally devised by his uncle; who, however, ultimately left it to his own brother Benjamin. The erection of this considerable addition to "the old house at home" clearly indicates a definite intention now to remain at Selborne.

To the Rev. John White.   Ringmer, Sepr. 11, 1777.

Dear Brother,—Being informed that Mrs. Snooke was seized with the palsy, and had lost the use of one side, and that her speech was much impaired; and moreover that she was alone by herself without any friend; I set out at a day's warning, though surrounded with workmen, and arrived here late last Saturday evening. I found the poor old lady in a low and languishing state, though better, the people about her told me, than she had been some days before. The next morning she was much mended; and has continued to mend so fast every day that she is become quite another woman; and Mr. Manning informed me this morning that he had now good hopes of a recovery.

Brother Harry brought your MS. to Selborne the first week in August but what between an hurry of business, company, and building, I have been able as yet to pay little attention to it. Yet though I have not paid it that regard which I ought, a visitor of mine has read it through with great care; and if I may judge from the many hours he bestowed on it each day should suppose he was well pleased. The person alluded to is Dr Chandler the traveller in Greece, who being no naturalist has no partiality for the Linn. system; but avers that it will prevent your book from becoming popular. He and I had much serious talk about the matter; and he asserts roundly, that he is *sure* that if you could perswade yourself to divest it of its quaint garb (those were his words) that he is certain it would be worth £200 of any body's money. He advises (no he does not, for he spoke with great modesty on the occasion) he hints, I should say, that if you could prevail on yourself to exchange *Classes* and *Ordines* for *Chapters*, and to throw all your tables back into an appendix, that your book would be very much read. The generality of readers, he observes, are very lazy, and afraid of figures; though your tables, he thinks, may be pleasing and useful to some. He farther added, that you might still refer to Linn. &c. at the bottom of each page. And I have observed myself, that booksellers lately in new editions of Nat. works have added Linn. names; and the reason is, because though it is the fashion now to despise Linn. yet many languish privately to understand his method.

Pray weigh seriously what I have said, and consider about the D^{rs} £200. You have not been informed, I think, that John Wells has at last consented to sell me the fields

Walker & Cockerell ph. sc

T. Holt White.

behind my house, that *angulus iste,* which the family have
so long desired. For this little farm I have laid down
some money in part payment; so hope no untoward accident
will now deprive me of it. With respects to my sister I
remain.   Yʳ affect. brother,
G. WHITE.

Pray write to Selborne where I hope to be soon.

His aunt's illness did not prevent his noticing, and
noting in his *Naturalist's Journal,* exact particulars
of the tortoise's diet on September 11th, and, a day
or two afterwards, the fact that it had not "at all
increased in weight since last year." On October 9th,
1777, Mulso wrote regretting that he could not visit
Selborne that year.

"I am angry with you that you speak so faintly about
your own work. Mind, that I expect you upon 'Nature,'
and the Bishop of London upon 'Isaiah and Prophecy'
next winter. Fail not herein 'as you shall etc.'"

In the following letter the elder brother records
his opinion of the MS. of 'Fauna Calpensis':—

To the Rev. John White.   Selborne, Octr. 31, 1777.

Dear Brother,—Had I not been called in the beginning of
this month to Oxford, where I spent all my time either in
college business, or inspecting, and transcribing by means of
an amanuensis, many curious papers from the Archives of
Magdalen College relative to the antiquities of Selborne, you
had heard from me some time ago. In my pursuits as an
antiquary Dʳ. Chandler has been wonderfully friendly and
communicative, and my discoveries about this place are very
great: we examined 366 parchments. I have now read your

work, all but the entomology, once over; and am proceeding to read several parts twice over. In the whole I *much approve* of your book. Your preface is neat; your history is what I call true Natural History, because it abounds with anecdote, and circumstance; and I verily think your dissertations on the *Hirundines* are the best tracts I ever saw of the kind, as they throw much light on the dark but curious business of migration; and possess such merit as alone might keep any book from sinking. If consulted I therefore *protest* loudly against the intention of throwing your papers aside; for I think in a thousand instances they will delight a good Naturalist. I therefore pronounce as the Vice-chancellor of Oxon. does on similar occasions—*imprimatur*. But then to act as an impartial critic, I must also say, that sometimes (and others think so as well as myself) your language is rather diffuse, and your sentences *too long*; and what I most wonder at is, that at times you not only use the same verb, or its derivations five or six times in a paragraph, but sometimes twice or thrice in the same sentence. Being jealous of the honor of your work, I cannot admit of these inaccuracies, and have therefore presumed to amend some of them, but with what success I must leave you to judge. I must therefore desire you, who are so perfectly capable, to bestow a fresh and severe inspection on the language. Brother Thomas is now in town, and I wish you would desire him to send me down your entomology which I long to see.

No wonder that you did not much relish D[r]. Chandler's proposal of rejecting all *system*; the reason of sending you that advice was that I thought *then* that *System* was the stumbling-block between you and your chapman; but *now* I plainly perceive that warm words and some heats have arisen between you, which I hope will all soon be forgotten. Indeed I wonder that in *these days* any work should stick on hand of your sort; as I cannot but think that it might sell. Would it not therefore be best to make fresh advances

in Fleet Street; and so set your work a going in some way?
When you print, pray correct the press yourself. Pray,
before every class, give an explanation of terms: Linn. does
so; and I think by this means the town might be led on
gently to relish Linn. terms. But without a glossary how
should men know what the *lorum** of a Bird is! no wonder
Linn. does not answer your letters; poor man, he is grown
childish!

Poor Nanny White was buried last Monday night in this
church-yard; she dyed at S. Lambeth.

If you lend money on private security, pray be careful.
Jack, I hope, will write to me about the earthquake.
Brother Thomas has the best interest with Mr. Lort; I have
none. Next week I put in my sashes, and proceed to ceiling
and plastering my great parlor. Our weather is very tem-
pestuous; the glass yesterday at 28·3. My best respects to
my sister.   Y^{rs} affect.,

GIL. WHITE.

On October 21st, 1777, Samuel Barker was admitted
a Pensioner of Clare Hall, Cambridge; as appears
from information kindly supplied by the Rev. Dr.
Atkinson, the present Master of that College.

To Samuel Barker.   Selborne, Novr. 7, 1777.

Dear Sam,—No event that I have met with for some time
has given me more pleasure than the news of your being
sent to the university: because, I trust, you will make the
best use of this advantage, both in your literary pursuits,
and by improvement in the knowledge of men and manners.
As to a proper acquaintance, you have nothing to do but to
lye by, and act a little on the reserve, and you will soon
discern what young men are suitable to your purpose: and

---

* *Lorum*, the space between the bill and eye of a bird.—A. N.

besides young people of your own turn, when they know you a little, will naturally make some advances.

All the house-martins withdrew about the 7th of Oct., and seemed gone to a bird 'til Novr. 4th, when 21 were seen playing about under the hanger all day, and for that day only. This circumstance seems the more odd, and amusing to me, because I have known it befal more than once or twice. Where were they during the interval? and where are they now? This event militates strongly in favour of hiding, and against migration. The bats do just the same all the winter and spring: they sleep at intervals; and then come forth and feed, and retire again.

The Order of *Polygamia frustranea* is constituted, you know, from having the florets of the disk hermaphrodite and those of the radius neuter. Not knowing where to apply for a common knap-weed in bloom, I know not how to solve your difficulty. The district round Cambridge will furnish you in the summer with the great aquatics. When you are a little at leisure I shall always be glad to hear from you.

Don't fail to practise frequently in writing English.

I am your affect^{te} friend,

GIL. WHITE.

In the previous month of October, 1777, the purchase of the "farm late John Well's," as he called it, was completed by Gilbert White. The money for its purchase was advanced by his brother Thomas, whose tenant he accordingly became at an annual rent.

On November 30th, 1777, Mulso writes :—

" I wish you joy of your purchases, of your buildings, and of the advances of Selbourne towards perfection. I feel a partiality for that place, from its being such a favourite

of yours, and from the many happy and useful hours that
I have spent there.

"I thank you for the piece of Mr. Grimm, but surely I
was never more disappointed. I declare that had the picture
come through any hands but a *White's*, which might have
directed me, I should not have guessed at the place. A
print in general does ill with perspective; but in this,
neither the Hill itself, or the neighbouring country are in
character. I hope I do not mortify you to say so: and I
hope better things of your other views."

The view in question was clearly a proof of the
vignette of the Hermitage. Perhaps the severe
opinion was partly justified, but it is not easy to
do justice to a scene on an eminence; as it were,
hanging in the air.

That Gilbert White was now fairly embarked in
his 'Antiquities of Selborne' appears from a long
letter, dated December 15th, 1777, about Selborne
Priory, to Mr. John Loveday, of Caversham, M.A.
of Magdalen College, Oxford, and the following :—

To Thomas White.
[With a list of documents from Dr. Chandler concerning
The Priory of Selborne].

30 Decr. [1777].

Dear Brother,—You see D^r Ch.[andler] has been wonderfully kind and obliging to take so much pains in transcribing
from the *index* such articles as he thought most interesting
with respect to the priory &c. of Selburn. From the *index*,
I say; for it doth not yet appear that he has at all consulted the original papers, which are reposited in the Tower
[of Magd: Coll:] under several keys. I have made application to the President by means of a friend for leave, if

possible, to be granted, to take *some* of these papers into my hands, for which I would give *bond* to the society. But the D^r says the statutes are so strict concerning these papers, that he doubts much whether my request will be attended to. All the indulgence I may expect to meet with, he imagines, if any at all, will be to have them *deposited in some Fellow's apartments,* in which I may have access to them. He desires much, that *no notice* may be taken of his having sent me this extract.

Pray read over these articles very *deliberately* and send me your sentiments. I think I discern many interesting anecdotes; but shall forbear to mention particulars, wishing rather to see our wits jump.   Your aff. Bro^r

GIL. WHITE.

I thank you for your letter this evening.

At the beginning of the year 1778 a bad account of John White's health reached Selborne from Blackburn. On January 5th, Mulso, after condoling with his friend upon the melancholy news, continues—

"You are a happy man who clear away rubbishes, and build on a clear surface. I shall visit your new room one day or another, I hope. With its beautiful site it will be one of the first rooms in the County.

"I have framed Mr. Grimm, though I dislike him as a Print. Where he could throw a little colour, or chiaro oscuro, the effect might be great, but Ned Mulso and Mr. Airson as well as myself, declared that they should never have thought of Selbourne from that Piece. However it will do with the rest. The lines are strong and clean, and poor Harry* makes a decent figure, but not so good as in the original. If I was with you, I could point out what would have been more advantageous; but the thing

* As the Hermit.

is set, and I do not desire to put you out of conceit with your Vignette, which is really pretty. But I grow very impatient for *the Work*, I have promised it as a regale to the good old bishop. I depend upon the religious turn that is in it to compleat his approbation to that part which as a naturalist he may know less of, and of course care less about."

On February 12th, 1778, the same correspondent writes :—

"You are so taken up as a builder, that you do not yet speak in your old style of a gardener. . . . The Hermitage is hanging over my chimney now, and I do all I can to persuade myself that it is like: but your little motto at the bottom does more towards bringing it to my mind than all Grimm's graving. Success to your *Lares*!"

The motto referred to is, of course, the line from the 'Invitation to Selborne'—

"Where the Hermit hangs his straw-clad cell."

A new correspondent, to whom gossiping letters were addressed, now appears in Mary (Molly), the only daughter of Thomas White, who, at this time a girl aged nineteen, kept house for her father, a widower, at South Lambeth.

The following letter was written after a visit to his brother at South Lambeth, whither he carried his *Naturalist's Journal*, in which he continued to note his observations, some of which read rather oddly at the present day.

" March 14. The green woodpecker laughs in the fields of Vauxhall.
" Owls hoot at Vauxhall."

To Miss White,
    Thomas White's, Esqre.,
        South Lambeth, Surry.

Selborne, Apr. 13, 1778.

Dear Niece,—It is now full time to return your father and you thanks for all good offices at S. Lambeth; and to remind you that you are indebted to me not only *for this year*, but *for all last year*, and I believe *all the year before*: so that I hope you will come when convenient and wipe out all scores.

My poor portmanteau-horse, Miller, has lost the use of his hinder parts; and now one of his legs is swelled to an enormous degree; so that he cannot lie down!

The *mint*-man, Charnley, has sent down my bed, table, &c. For the table he has charged £2 8s. 0d., whereas a man in Fleet-ditch, only three years ago charged but £1 14s. 0d. for the *same* sort of table, exactly of the same *dimensions* and *materials*; and a very *well-made* table it is, and very handsome.

Please to tell your father that his Greatham tenant has just paid in an year's rent up to Michaelmas last.

Poor Mr. Willis of Holiburn is ill, and gone to Bath. The weather is so hot, that we all say there never was such weather before; forgetting that the 26th and 27th of March, .77, were much hotter. In my way down I found a swallow at Ripley: and we have here swallows, and cuckows, and nightingales!

Your handmaid, I hope, arrived safe, and proves to your mind. I address you now as a prudent and experienced mistress of a family. By a late letter brother John rather mends; but his poor wife is worn down, and very low indeed. Did you say how does my new parlor go on? Why, pretty well, I thank you: and was it never to rain any more would be very dry: however this weather acts

much in its favour. I* . . . . . begin to alter Thomas's
room, and to make my entry* . . . . . and then shall
proceed to flooring, &c. I am now going to retain my
weeding woman for the summer. This is the person that
Thomas says he likes as well as a man: and indeed ex-
cepting that she wears petticoats, and now and then has
a child, you would think her a man. To the care and
abilities of this Lady I shall entrust my garden, that it
may be neat and tidy when you come. My great parlor
grate and fender are arrived, and seem proper for the
place.

My cucumber-plants are gross, and vigorous; and I have
one, and only one fruit about the size of my thumb. As
the season was late I just saw my crocus's in bloom. On
Thursday morning last we had thunder and an heavy
shower. I have seldom in this month seen the weather
so hot and the ground so moist at the same time; of course
therefore vegetation is very vigorous. When you can spare
time from the cares of housekeeping, and want to relax your
mind, I shall be glad to hear from you, and shall rejoice
in your communications.

I am, dear Niece, y^{rs} affectly,

GIL. WHITE.

Respects as due.

I have just received a letter from Ringmer. Mrs. Snooke
is much mended.

To Mrs. John White.   Selborne, April 17 [1778].

Dear Sister,—By both your last letters, for which I
return you thanks, it plainly appears that my brother continues gradually to recover strength, and that air, exercise,
and bathing are of singular service; and therefore I hope he
will strive against irresolution, and summon up all his man-

* Letter imperfect.

hood to pursue the one and submit to the other, irksome
as it may feel at times.   You talk of Bath in this case:
and those waters doubtless have done wonders ; but brother
Thomas says while the *cold* bath continues to be so service-
able, he cannot see what more can be expected from *hot* ones,
which, one should suppose, would rather relax.   He thinks
at present you had better pursue your home regimen.   In
town I saw Mr. Fielden, and your intended curate ; the
former had lately seen my brother, who to his thinking was
marvellously mended, and looked in the face almost as
usual.   Yesterday, if I mistake not, Mrs. Snooke entered
into her 84th year.   The late hot weather was of singular
service to her, and relieved her from a cough, which had
annoyed her the winter thro'.   On Easter Monday brother
and sister Harry and several of their children are to go up
to South Lambeth.   They have just inoculated four of their
children with singular success.   My neighbour Yalden has
just got a regular smart fit of the gout.

My new parlor now dries at a great rate ; and will be fit
for use at Midsum^r, but I shall not be able to compleat it
this summer.   I must not put on my upper paper 'til
another year.   With my best wishes and prayers for my
brother's recovery, I remain

Y^r affectionate brother,

GIL. WHITE.

CHAPTER II.

On April 24th, 1778, the *Naturalist's Journal* records the discovery, in bloom, "in the Litton coppice at Selborne just below the church," of the "*Lathræa squammaria*, a rare plant," and a little later a slip is inserted noting that his servant Thomas Hoar had heard "pretty late one evening" the twittering notes of swallows from "under the eaves of my brewhouse, between the ceiling and the thatch."

· "Now the quære is, whether those birds had harboured there the winter through, and were just awakening from their slumbers; or whether they had only just taken possession of that place unnoticed."

A similar occurrence, reported by Mr. Derham to the Royal Society, is referred to.

"July 3. Began to inhabit my new parlor."

This room was built at the west end of the old house, looking into the garden, and "outlet," towards the Hanger, by two windows. In recent years a story has been added above it, and the windows altered, while a passage to a new room now occupies part of the original room, opposite the windows.

To Mrs. Barker.

Selborne, Sep. 2, 1778.

Dear Sister,—My thanks are due for your kind letter. I have now the pleasure of seeing my house full of friends. My niece Anne Barker pleases me much, and is a sensible intelligent young woman. Mrs. K. Isaac has not been here for 25 years; and finding every thing much altered, hardly knows the place again. Molly White and your daughter seem well pleased to meet again. Jenny and Becky White* are to come to Newton this week: Mrs. Yalden pressed their mother much to come; but she is in a very poor way, and chose to wave the journey. Mrs. Snooke has just written to me with her own hand: she did not complain much. Last post I had a letter from Blackburn: my brother's state of health and spirits is much the same: my poor sister makes sad complaint, and laments the state of their family: indeed they both merit the compassion of their friends. My great parlor is now of singular service; but while it is so empty the echo is very troublesome. I have a new bed in my little red room; and have put my old white bed up in my late drawing-room,† where I lie, as you ordered me. Brother Harry's school thrives: he has just got three new pupils, and expects one more. His house is now quite full. My peaches ripen; but, the summer considered, are not so fine as might be expected. We have fine wheat; and a vast crop of hops. Barley and oats are lean and poor. The failure of turnips is miserable!

Your loving brother,

Gil. White.

* Daughters of Benjamin White.
† This room was over the kitchen on the first floor, looking into the garden. Mr. Bell notes that in this room the Naturalist died.

THE OLD PARLOUR AT "THE WAKES"

(This room is quite unaltered since Gilbert White's time)

[To face p. 28, Vol. II.

To Samuel Barker.

[On the same sheet as the above.]

Dear Sam,—I am much pleased to find that the university, and your studies there give you so much satisfaction: there is no fear that you will neglect this opportunity of improvement, or spend your time amiss: I should rather wish that you were cautioned to remember, that it is possible for a young man to apply too earnestly; and therefore I hope you will intermix daily exercise with your studies. To the generality of young men I am well aware this caution would be needless; but to you, who I know apply yourself to all laudable pursuits with all your might, it may not be improper.

D^r. Chandler the traveller has been with me a month, and is just gone: he has furnished me with *more curious matter* respecting the antiquities of this place; and in particular with William of Wyckhame's *Notabilis visitatio* of this priory. From this long instrument, consisting of 36 injunctions, and reprimands, it appears, that this institution, which then had been founded only one century, had deviated much from its original simplicity. For they were become mighty hunters, and used to attend junketings, and feastings; had altered their mode of dress; and used to let *suspectae* come into their cloisters after it was dark; had suffered their buildings to dilapidate; had pawned their plate; administered the sacrament with such nasty cups, and musty, sour wine, that men abhorred the sight (*ut sit hominibus horrori*) had let down their number of brethren from 14, the original number, to eleven; had suffered their friends, and relations to hang on to the convent, and eat it up, &c.: they also were got into a method of laying naked in bed without their breeches, for which they are much reprimanded. Moreover I find that the Knights Templars by their statutes were enjoined constantly to sleep in their

breeches, and to have candles constantly burning in their
dormitory. Should I ever be able to finish my work
respecting this my native place, the old deeds and charters,
&c. will furnish a long appendix.

Your affectionate friend,

GIL. WHITE.

Towards the end of September the annual visit
was made to Mrs. Snooke at Ringmer, where
Timothy, the tortoise, was noticed, and, as usual,
weighed. Returning by Findon and Chilgrove,
the *Naturalist's Journal* records—

" Oct. 8. [Findon.] Not one wheatear to be seen on all
the downs. Swallows abound between Brighthelmstone and
Beeding. Not one ring-ouzel to be seen on the downs either
coming or going.
"[Oct.] 9. [Chilgrove.] Many martins near Houghton-
bridge. Some swallows all the way."

On arriving at Selborne, the *Naturalist's Journal*
records a sight which tells indeed of the difference
which modern farming and game preserving have
caused in wild-bird life.

" Oct. 13. Near 40 ravens have been playing about over
the hanger all day."

On October 15th Mulso wrote that he was unable
to accept an invitation to Selborne, though passing
through Alton to Farnham Castle.

" I looked up towards your hills as I passed them with a
longing eye, and I passed on without the unfeelingness of the
Levite. . . . I hope your excursion has been of service to
you, and that you can sleep without dreaming of the French.
Mrs. Snooke is not so fainthearted, or she would not hold so

well at her age. I hear of accommodations, but I trust no
reports; at the same time I am not apt to fear them. Wrap
up your content in the conclusion of Voltaire's *Candide*,
' Il faut cultiver notre jardin '!"
"We met sister Chapone at the Castle, who helped to
enliven the place."

To Miss White.
Selborne, Oct. 19, 1778.

Dear Molly,—Bating your account of your father's indisposition, which I hope will be very short, your letter was
very agreeable to me; and particularly the circumstance
which intimates your intention of coming down next
Monday. Nothing, I hope, will prevent so agreeable an
event; and I will take care to send Thomas in time on that
day to meet you at the inn at Chawton.

For some mornings past we have experienced severe frosts
for the time of year, which have stripped my vines of all
their leaves, and left a fine crop of grapes naked, and
forlorn on the walls; they used to be cloathed with foliage
'til the middle of next month! So you must come and eat
grapes every day, or they will be spoiled.

A Selborne man was aboard the Porcupine sloop when
she took the French India rich ship. I saw a letter from
him this morning, in which he says that his share will come
to £300. This will be some recompence to the poor fellow,
who was kid-napped in an ale-house at Botley by a press-
gang, as he was refreshing himself in a journey to this place.
The young man was bred a carter, and never had any
connection with sea-affairs.

Your hand-writing is very fair, and handsome: pray keep
it up; and don't scribble it away. Nanny Barker is a
very good correspondent; but spoils her hand by writing too
fast. A little patience would make her also a good pen-
woman.

Mr. Etty, who is going suddenly to town, will take my letter with him. Mrs. Etty is well. I am with due affection,

Y^r loving Uncle,

GIL. WHITE.

I do not recollect any more errands. Respects over the way.*

To Mrs. John White.   Selborne, Nov. 2nd, 1778.

Dear Madam,—I thank you for your last letter, which gave a much more favourable account of my brother's health than any that I have received for some time; and may, I hope, be followed by many more to the same purpose. You may tell my brother that Dr. Chandler has read over *every part* of his work with great attention; and approves of the whole much, and was much entertained with many parts: but does not, as my brother knew before, relish the systematic manner in which it is drawn up.† He has in several parts with his pencil altered several expressions, but chiefly where the same verbs, etc., are used two or three times in a sentence: such slips must necessarily befall "opere in longo": with the matter he has never meddled. I have also read over said work with great care (the insects very lately) and approve much of the whole, which discovers, I think, great discernment, and application. Here and there I have flung in a small marginal note. Many parts are to me curious, and interesting: and the whole Fauna contains much more anecdote than ever I met with before in such a work. Some parts are, and must be in so long a work, less engaging than others. The Hawks, the *Hirundines,* the *Turdi,* the *Gallinæ,* the Insects, are favourites with me: not

---

* Benjamin White lived opposite his brother at S. Lambeth.

† The Linnean method was for many years unpopular in this country—a fact which perhaps explains Benjamin White's refusal of his brother's work, and the consequent abandonment of its publication.

but the other *Ordines* have each their merits: but one man is pleased with one subject, and one with an other.

I wrote to Nephew Benjamin last Saturday, and made the proposal mentioned to me. But I would have my brother at present sit loose to such matters, and not let his mind be agitated about this event, or any other; but keep himself as quiet as possible. As for the work I could wish to see it published.

The present new Lord Chancellor * has given a decree point blank against us with respect to Mr. Holt's concerns. Mrs. Ben. White continues still in a bad state. My brother Thomas and Molly are just returned to my house: my brother has been ill; but is recovered. I have sent word to Harry about his rents. You did right, I think, in allowing the repairs.

Your loving brother,

GIL. WHITE.

When you see or consult your physician at Manchester, should he not be asked whether he has any opinion of electrical applications? My brother formerly used to amuse himself with electricity.

The antiquities of Selborne were now engrossing their historian's attention. His interest in these, however, by no means extended to his friend Mulso at Meonstoke, who wrote on February 13th, 1779 :—

"You know, wretch, that I have always had, and still have such an opinion of your precision and integrity, that I proclaim things, *as certain*, that you have once said. Take care that you prove well what you say of birds of passage, of spiders and flying webs, for I shall assert it 'pedibus manibus-que' on your authority. I am shocked at you for deferring that Piece so long: for heaven's sake do not take too much

* Lord Thurlow.

time in ascertaining the size, the markets, the tolls, the souls, the priories, and religious Houses of Selborne; for these circumstances, though curious in reality, are to the *goût* of not five readers in five hundred. Be it therefore very clear, but very *short*. The novelty and elegance, the tenderness, and the *piety* of the natural part will be the forte of the Performance. . . . How was it with Mrs Chapone; it was the genuine affetuoso, the *con amore* of her book that gave it its run. . . . Pray come out while the passion rages; the world is getting off its eyes from Portsmouth and the Trial.* . . . Now's your time."

Shortly after this time the brothers at South Lambeth received a visit from Gilbert White, who found Benjamin lately a widower.

To Miss White.   Selborne, April 17, 1779.

Dear Molly,—My thanks are due for your kind letter, and for your father's care in procuring me two fine hams; and for his present of a rain-measurer; and for his trouble in purchasing my long annuity. There *was* a time when rain-measurers were very entertaining; and doubtless there *will* again: but now we have seen no rain for four months! I rejoice much to hear that my nephew Thomas recovers so fast. Enclosed I send some large white cucumber-seeds for your father: but the sun is so hot, and the dung so dry, that hotbeds thrive but poorly. When Mr. Cricket is tired of his new room, he may let it to you. My furniture from Mr. Graham does not come to Alton 'til this day. As soon as I can get a person I shall paint my room. Mrs. Rashleigh left Selborne this morning. Old George Tanner

---

* Admiral the Hon. Augustus Keppel was tried by court-martial at Portsmouth, January 7th–February 11th, 1779, and honourably acquitted of a charge of cowardice. Politics had a good deal to do with the trial, and party feeling at the time seems to have been greatly excited.

lies very ill, and is in danger. In the night between the
11th and 12th of this month Burbey's shop was attempted.
The assailants wrenched off the hinges at the bottom of
the shutters, and so crept up between the shutters and
the sash, and broke the glass; and with a knife began to
cut and hack, in a very bungling way, the bars of the
sash. It is imagined that they were disturbed in their
business by some means, for they never got in, nor could
reach to the *till*, which is near the window, and was, it
is supposed, the object they had in view. Burbey heard
the glass jingle; but being but half awake, did not know
what was the matter. Our maypole is mended and painted:
we talk of gilding the Vane. If I had not interposed, the
vane would have rested again on a *shoulder*; but now it
is to turn on a pivot at the top.*

As soon as ever you hear about Molly Barker's finger,
pray give me a line.

> With proper respects I remain,
>> Your loving uncle,
>>> GIL. WHITE.

The French East India-man, the Carnatic, met with such
stormy weather to the E. of the Cape of Good-hope that
she could not get to that port at all; but arrived at last
in a most distressful condition at the Isle of Loanda on
the coast of Angola, being in about Lat. 8. S. The Por-
tuguese Governor at first pleaded an inability of assisting
them, saying that they had been without rain for two
years, and were almost starved: but the captain urging
their great wants and pitiable condition, and adding that
they had a sick English lady aboard, they were at last
admitted to go ashore; and miss Shutter† was lodged in the
Governor's house, where she was entertained for a fortnight.

---

* The maypole stood on the Plestor. It is clearly seen in the pictures
given in the quarto editions of 'The Natural History and Antiquities of
Selborne.'

† Mrs. Etty's niece, who had recently returned from Madras.

On July 22nd, 1779, Mulso writes :—

"You will be very happy in the company of your brother
Thomas and his daughter. I am not delighted at present,
though I know not what I may be, at your labours about
the history of Selbourne. I fear the sweet and elegant
simplicity of your observations will be overwhelmed by
the rubbish of the antiquities of your native place. *I* shall
be pleased from the partiality I have for the place for
your sake. The Provost of Worcester,* and some of your
antiquarian friends will like it for the studiousness of the
researches; but I doubt whether the book will be the
better for it in the eye of the world."

Then he goes on to an observation, which now, so
long after the fame of Selborne and its historian
has become established, seems a singularly prophetic
one :—

"It may save some future biographers trouble, who may
think it necessary to celebrate the place, where such a
genius was born. . . . I saw Mr. Wyndham lately: he told
me that he had hopes to have seen you, while Grimm was
with him; and that he had been surprized and delighted
with the grandeur of Selbourne Hangers."

On September 27th, 1779, Mulso writes again :—

"I called on Mr. Buller at Alresford, and he told me of
your having in your option the living that you had long
had in your eye. We wondered whether you would resolve
upon taking it or no. I own I should think you very
wrong if you did not. You will be money out of pocket
for a year or two, but you will be repaid hereafter. The
situation and the distance are both of them strong tempta-
tions and really good circumstances. The farmers cannot

---

* Dr. Sheffield, a friend and correspondent of Gilbert White.

but expect a rise : you are in the right not to think of straining them, but you have prudence enough not to say so . . . at all events you will raise your living to something more than it stands now, as Mr. Cowper was on it a great number of years at the old rent. The curate there is a valuable acquisition, and now I hope to see you master of your own time. . . . I wish you a good journey to Sussex; I fancy you will find there a strong persuasive to taking Ufton."

The last sentence, in the light of what follows later, clearly means that her nephew would find Mrs. Snooke, from whom he had some expectations, in good health.

The living in question, which Gilbert White visited on September 10th, was the rectory of Ufton Nervett, in Berkshire, in the gift of Oriel College, which lay conveniently enough for an in- cumbent who might be partly resident at Selborne, since it is only seven or eight miles north of Basing- stoke ; but it was declined after being again visited in November.

Mulso wrote again on this matter on December 21st, 1779 :—

"I cannot but approve of your refusing Ufton upon the reasons that you give. A living is a very troublesome charge, and there are but two reasons for burthening oneself with it, 'the hope of doing real good,' and 'the reasonable expecta- tion of a large increase of income.' The first you could have done as well as any man, had you chosen a constant resi- dence there; but yet there does not lie so much spiritual power and efficacy in the clergy of the Church of England

now, as did formerly. The itching ears even of the vulgar; and
the republican principles of the Times, make all the members
of our church looked upon with an evil eye. As to the
last you are the best judge of it; but in my opinion, a certain
small income is better than a precarious large benefice."

In the following letter a new correspondent
appears, the Rev. Ralph Churton, who became one
of Gilbert White's most intimate friends and visitors,
until the latter's death. At this time a Fellow of
Brasenose College, Oxford, and twenty-five years
of age, he subsequently became Bampton Lecturer
(1785), Whitehall Preacher (1792), Rector of Middleton Cheney, Northants (1792), and Archdeacon
of S. David's (1805). Like his Selborne correspondent, he was a friend of Mr. John Loveday,
of Caversham, father of John Loveday, D.C.L., of
Williamscote, Oxfordshire ; and of Dr. Richard
Chandler, the antiquary and traveller. He wrote
in later life numerous theological works.

To the Rev. R. Churton.   Selborne, Nov. 17, 1779.

Dear Sir,—On opening your favour, I was much pleased
to see your name at the bottom ; because you are a gentleman to whom I am much obliged, and to whom I wished for
an occasion to express my acknowledgements.

You are a fellow of a college as well as myself, and therefore must be well aware that with regard to elections it
is not in my power to enter into any promises ; but you may
be well assured that I shall have the better opinion of Mr.
Smith for what you say of him, and, if I am able to attend
at Easter, shall mention your recommendation to the society.

When the summer is established, if you find within yourself an inclination to visit Hants, I shall be very glad to see you at my house, and to show you our prospects, which are romantic enough. Your company and conversation, provided you can bear with the infirmities of a deaf man, will be very agreeable to me. D^r. Chandler is now sitting at my elbow, and is deeply engaged in Bishop Waynflete's Registers, two volumes folio which I obtained to be sent to my house from Winton by permission from the Bishop of that diocese; last summer we had Bishop Wyckhame's registry of the same bulk and number of volumes. I am, Sir,

Your obliged servant,
GIL. WHITE.

So far back as September, 1774, in Letter XXII. to Barrington, the Naturalist had complained of deafness as partially disqualifying him from observing nature. Among his effects at his decease was an ear-trumpet. That however he continued to enjoy good health in other respects appears from a sentence in a letter of Mulso's, dated August 16th, 1780 :—

" You have owned yourself threescore with only one Infirmity."

To Miss White.   Selborne, Decr. 4, 1779.

Dear Niece,—When I wrote last I was desirous to wait on you and your father as next week, but the difficulty of getting my church supplyed on these dark, short Sundays; and the nearness of Xmass, against which I must be back at all events, have abated my ardor; so that now I think it best to defer my visit 'til after the holidays. I am very sorry indeed to hear that your father has experienced some return of his fever; and sincerely wish the medicines he is taking may have their effect. Many thanks for your kind letter, which was opened at Mr. Etty's, where we all admired the neatness of your hand, and the propriety of your words; and agreed all that you were a very nice maiden.

Pray write again soon, and let me hear, if it please God, that your father is quite recovered.

Mrs. Etty was this morning at Hartley, and paid a visit to Mrs. Wilmot. Mr. E. and I made an evening visit about a week ago. Mrs. W. seems to be an accomplished woman; and had, I imagine, a large landed fortune. There is a large young family; two boys at school; Miss W. a young lady in person not unlike yourself; and three little Missisipies all in a row. They buryed a daughter about a year ago who was near grown to woman's estate. We have had vast rains lately.

I am your affectionate Uncle,

Rain.
GIL. WHITE.

Decr. 2, 1·40.

„　3,　·60.

Pray tell my brother Ben. that I wish he would send me down Harmer's 'Observations on the usages and customs of the East,' and a small pocket-clasped almanack.

To Miss White.　　Seleburne, Dec. 16, 1779.

Dear Molly,—It will be by no means proper to send you three or four cheeses from hence; because the cargo* at our shop turns out very poor, and mean, without any good flavour, and full of eyes; so that I hardly can pick out a tolerable one for my own table. There is a Randal at Farnham, a cheese-monger of repute; would it be worth while for your father to write to him?

---

* This word, in the sense of a parcel conveyed by land, has long become obsolete. Wykehamists, however, still use it of a present in kind from home.

We were all pleased to hear that your father was so well recovered: when he gets about, and goes to town, I wish he would send me down half an hundred of good salt-fish. There were vast rains with much thunder and hail on Monday last: so that I fear the water will encrease in your *hold*, and that pumping will not avail long. The springs at Faringdon rise very fast, but are not yet up to the surface.

Mr. and Mrs. Wilmot, Miss Wilmot, Miss Fletcher, and Captain Bain were to have drank tea at the vicarage last night; but were prevented by the snow.

On Friday, I hear, the poll-books, and all election implements are to be wafted over to the Isle of Wight, which is the stronghold of the Baronet candidate. At present Mr. Jervoise is ahead. This day Mr. Etty is to meet his brother at Alresford.

With due respects I conclude,

Y^r affect. friend,

GIL. WHITE.

On March 8th, 1780, Mrs. Snooke died at Ringmer, aged eighty-five. Her nephew attended her funeral there on the 15th, and thence wrote :—

To Benjamin White.   Ringmer, Mar. 16, 1780.

Dear Brother,—After returning my sincere thanks for all the good offices that I have experienced from you and yours so lately; I think it proper to inform you, that Mrs. Snooke by will has given me her Iping-farm,* charging it with a legacy of £50 to you, and £50 more to be divided equally among all your children; and also with £50 a piece to each of her nephews and nieces. These bequests she has enjoined me to pay within twelve months after her death.

* Which had belonged to her father, the Vicar of Selborne.

I shall say nothing concerning the other parts of her
will, because Brother H[enry] in his letter has entered into
particulars.

In my journey I have caught a cold, and cough; and am
feverish; so that I shall be glad when I am got home.

I propose to leave this place on Friday, and to return
by Uckfield, Cuckfield, Horseham, Dorking, Guildford, &c.:
the country as far as Guildford will be new to me.

With due respects I remain

Your loving brother,

GIL. WHITE.

The *Naturalist's Journal* records—

"March 17. Brought away Mrs. Snooke's old tortoise,
Timothy, which she valued much, and had treated kindly
for near 40 years. When dug out of its hybernaculum it
resented the Insult by hissing."

Frequent mention of the proceedings of the tortoise,
its weight, habits, food, etc., appear from this time
in the *Naturalist's Journal*.

On March 18th Mulso wrote :—

"I enter into your feelings at quitting Ringmer, a neat
and beautiful spot, and never entered without being associated
with the idea of a warm and valuable relation, and hospitable hostess. I know nothing of the value of the farm
that your aunt has bequeathed you, or of its condition;
but have hope from your silence on that head that there
is not such in it as would vacate your fellowship at Oriel
College—a circumstance which I touched upon lightly to
you of late, when you sent me word of your refusing the
living, when without much merit of a divining spirit, I
foretold the death of your aunt. I am glad to hear that

Harry will be materially benefited by her will, as he has a large family and is of our trade, which is not a very thriving one as times go."

There is something perhaps a little incongruous in the anxiety lest his friend should continue to hold his Fellowship without warrant, coming as it did from a man who at the time was a much (though legally) beneficed pluralist. It will, nevertheless, be well to endeavour to ascertain whether or no Gilbert White was justified in continuing to retain his Fellowship.

The original statutes of Oriel College, of January 21st, 1325–6, provide for the avoidance of a Fellowship on the obtaining of a " uberius beneficium." By an ordinance of 1441, made by the College in pursuance of its power under the original statutes and duly confirmed by the Visitor, it was provided that any of the Fellows who obtained "aliquod beneficium, redditus, patrimonium, officium, pensionem, seu pluralitatem eorundem, seu aliquam aliam promotionem, quocunque nomine censeatur, unde potuerit ad suam exhibitionem [*i.e.* support] annuatim, ad terminum vitæ, decem marcas de importatis percipere Oxoniæ residendo, eo facto, et alias per acceptionem hujusmodi promotionis," should be deprived of his Fellowship.

This standard of ten marks became inapplicable after the change in the value of money, but no other sum was ever authoritatively substituted for it. It

appears that the practice in the eighteenth century
was to take the then value of a Fellowship from
all sources, including allowances during residence, on
an average of the last seven years; and if the value
so ascertained was exceeded by the value of the
benefice or other property, the Fellowship was held to
be vacated. But the only cases which are noticed in
the College register are those of ecclesiastical preferment; and the value of these was taken from the
King's Books. It was not till 1816 that the College
confirmed this practice by a College order.

The above account is kindly furnished from Oriel
College; but the opinion of other experts in the history of College statutes, and their interpretation by
college custom (which is quite as important as the
statutes themselves), differs widely as to what is meant
by the words "patrimonium" and "redditus"—
whether the former word was held to include *all*
inherited property, or only patrimony in its strictest
sense; and as to what was held to constitute "redditus," or income; some authorities holding that
nothing but ecclesiastical preferment or distinct
"patrimonium," *i.e.* inheritance of real property,
would void a Fellowship. Certainly income earned
by a Fellow's own exertions, including a stipend
received as curate, has never been held to disqualify
him; and it would seem naturally to follow that
money saved by a Fellow during his tenure of a
Fellowship would not be reckoned as disqualifying
him.

Now Gilbert White never held any benefice other than his little living of Moreton Pinkney, the value of which, as reckoned in the King's Books, would be insignificant indeed. What, then, was the income from his total inherited property after his aunt's death?

Though he made fairly complete entries of his expenditure in his account-books, he does not appear to have kept any regular account of his incomings. Fortunately, however, the following entries occur of the rent and outgoings of his landed property, all of which was certainly inherited by him :—

YEARLY RENT AND OUTGOINGS IN 1780 OF FARMS BELONGING TO GILBERT WHITE.

	Rent.			Outgoings (1 year).			
	£	s.	d.		£	s.	d.
Holtham farm . .	15	0	0	Land-tax and quit-rent	2	5	6
Parson's farm [Sparrows Hanger] .	13	0	0	Do. do. .	2	19	6
Benham's land, Wakes [i.e. land near his house] .	8	8	0	Do. do. .	2	17	3½
Woodhouse farm [Sussex] . .	60	0	0	Do. do. .	10	15	11½
Hurst farm, Iping [bequeathed by Mrs. Snooke] .	39	0	0	Land-tax . .	6	10	9
(rent from Michs., 1780)							
Total rent . . .	135	8	0	Total outgoings .	25	9	0
Deduct outgoings .	25	9	0				
Net rent . . .	109	19	0				

From the net rent of Hurst farm, bequeathed by Mrs. Snooke, however, is to be deducted the interest

of the sum of £50 bequeathed by her to her nephew
Benjamin, to his children £50, and £50 apiece to
her nephews and nieces, Thomas and John White,
Basil Cane, Mrs. Barker, and Catherine S. Isaac; in
all, a sum of £350, which, at (say) 5 per cent. = £17
10s. per annum, leaving £92 9s. 0d. as the net income from landed property.

But, as all landowners know very well, and indeed
as appears from Gilbert White's accounts, there are
other deductions besides land-tax and quit-rent to
be made from the rent of land, such as insurance and
repairs of buildings, allowances to tenants, etc.

As regards personal property Gilbert White had
inherited and received a sum of £300, in 1746, on the
death of his great-uncle, Mr. Thomas Holt. A good
"economist," as his friend Mulso termed him, it had
been his habit to put by money all his life, which he
invested from time to time in the purchase of "Long
Annuities"; but, excluding this property, which, with
the exception of the £300 bequeathed him by Mr.
Holt, appears to have been entirely his own savings;
his income from all inherited property, after necessary
outgoings were deducted, can have little, if at all,
exceeded £100 per annum at this date (1780).

It remains to ascertain the value (including
allowances) of his Fellowship. In some of his
earlier account-books the amount received by him
"From my Fellowship of Oriel Coll:" is occasionally
entered.

	£	s.	d.
Thus, in 1755, he enters . .	50	14	4
,, 1759 ,, . . .	65	15	0
,, 1760 ,, . . .	116	11	10
,, 1761 ,, . . .	74	11	6

From the latter date no entry occurs until 1781, when he enters—

	£	s.	d.
"By drt. from Oriel College	146	0	0 "

an amount to which something should be added for the "allowances" appertaining to a Fellowship. The next entry, in 1787, is £154 15s. 11½d. Where, then, does the alleged impropriety of Gilbert White's retention of his Fellowship appear, and what is to be said of the statement already quoted, that he continued to hold his Fellowship by holding his tongue?

It is naturally very difficult so long after his death to put forward positive and absolutely conclusive evidence that the Naturalist was justified in retaining his Fellowship, of which no doubt the College would not have thought of depriving him, or any other Fellow, if his property had produced an income only a little in excess of the value of the Fellowship; but it may certainly be said that, apart from the abundant proof of the respect and regard which Gilbert White received from all who knew him, the whole evidence now procurable acquits him absolutely of a charge which should never have been so lightly brought forward against the

memory of a dead man : a charge made, it is true,
with very imperfect knowledge of the facts of the
case, though this circumstance hardly tends towards
its excuse.

To Miss White.   Selborne, Mar. 31, 1780.

Dear Molly,—It gave me much concern to hear that your
father had experienced some return of his complaint; but
I trust that the bark, and the advance of summer weather
will soon restore him to his usual state of health.   I hardly
ever remember an ague in this village, that properly belonged to the spot.   Charles Etty brought one lately from
school; but has known no return since his first arrival.
If your father has any idea that a change of air by and
by might have a good influence on his health, you may
assure him that I should be glad to see him here; because
nothing would please me better than any expedient that
might probably contribute to his welfare.

Timothy the tortoise accompanyed me from Ringmer.
A jumble of 81 miles awakened him so thoroughly, that
the morning I turned him out into the garden, he walked
twice the whole length of it, to take a survey of the new
premises; but in the evening he retired under the mould,
and is lost since in the most profound slumbers; and
probably may not come forth for these ten days or fortnight.

Mrs. Etty is very well, and so is Miss E. but Mrs.
Yalden is troubled with pains in her limbs.   Charles sails
for India about the end of April.   Pray write to me soon.

With due respects I remain,
Y^r affect. friend,
Gil. White.

On Monday I paint the great parlor.

Has your father been so kind as to receive my long ann.
up to Xmass last ?

To the Rev. R. Churton.

Selborne, near Alton, Hants.

July 3, 1780.

Dear Sir,—As I have always wished to express my gratitude for the many good offices you have conferred on me, I
must desire that you would furnish me with an opportunity,
by taking the trouble to come to my house, where I shall
rejoice to see you in the course of this summer.

At present my beds are all like to be full for two or three
weeks to come; but by the end of July at farthest I shall
be glad to see you for four or five weeks. It will probably
be in my power to shew you a new country, and a district
not unpleasing in fine weather. If you can bear with the
infirmities of a deaf man, your company and conversation
will be very agreeable to me; and in your answer I bar all
proposals respecting some future summer, because at my
time of life there is little dependence to be made on distant
engagements. Pray take me, in the very words of Creech,*
"just as I am, very much disposed to receive you, and ready
to show you all civilities."

If you are a botanist, we have a very good *Flora*, to whom
I am willing to introduce you. You are, I find, learned in
yew-trees: we have at hand several noble ones.

We have just found a large stone-urn down at the Priory;
for what use it was made it remains for you to inform us.†

We will examine The Temple, King John's hill, &c. &c.

I am, with great esteem,

Yʳ obliged, and humble servant,

GIL. WHITE.

* Thomas Creech (1659-1701), the translator of Lucretius, Horace, etc.

† *Vide* 'The Antiquities of Selborne,' Letter XXVI., in which this discovery is recorded as occurring "two years ago." The "judicious antiquary,"
mentioned in a note as guessing that the vase might have been a standard
measure, was no doubt Dr. Chandler, who, as appears from the postscript to
this letter, was at Selborne at the time.

Dᵣ. Chandler, who is going to be very busy with Bishop
Beaufort's Register, from Winchester, joins in respects.
When my beds are at liberty I will write: pray let me hear
soon.

From the *Naturalist's Journal*—

"July 1. We put Timothy into a tub of water, and found
that he sank gradually, and walked on the bottom of the
tub; he seemed quite out of his element and was much
dismayed. This species seems not at all amphibious.
Timothy seems to be the *Testudo græca* of Linnæus. Dᵣ.
Chandler, who saw the operation, says there is a species of
tortoise in the Levant* that at times frequents ponds and
lakes; and my Bro[ther] John White affirms the same of
a sort in Andalusia.

"[July] 11. Finished my great parlor, by hanging curtains
and fixing the looking-glass."

The "great parlour"—a name which, it should
perhaps be said, was at this time a recognised one for
a certain kind of sitting-room, and does not mean
merely "a large sitting-room"—being at last finished,
some little account of its construction and furnishing
may be of interest, and is obtainable from a bundle
of bills neatly tied up and endorsed by Gilbert
White, "Bills relating to the building of my great
parlor; in 1777, and furnishing."

Of course, no contractor was employed, and the
labourers' wages appear in their accounts. One
George Kemp, whose descendants are living in
Selborne at this day, was the foreman bricklayer, and

* The freshwater tortoise (*Emys orbicularis*) is found throughout the
greater part of Europe, and formerly inhabited England.—A. N.

"THE WAKES," SELBORNE, GARDEN FRONT

(The two windows on the left of the picture are those of the "great parlour")

[To face p. 50, Vol. II.

his wages were 2s. a day, his assistant receiving
1s. 6d. "Building bricks" cost 16s. 10½d. a thousand. "Rubling bricks from Harting comb" were
3s. a hundred. The carpenters employed were paid
1s. 8d. a day, and some of the nails they used,
presumably made by hand, cost no less than 1s. 8d.
a pound. Most of the timber used was brought over
from Winchester. A chimney-piece, described in the
bill as "23 foot 7 in. of superfishal white and vained
Italian marble" (£5 17s. 11d.), was set up in July,
1778 ; and a "large fine bath stove grate" and fender
(£4 9s. 0d.) were added a little later.

Hanging the great parlour with "a flock sattin
paper" cost a good deal more than would now be
paid, viz. £9 15s. 0d. A looking-glass, no doubt a
pier-glass, was bought in London for £9 19s. 0d.,
a price which, unless it was a very large one indeed,
may seem high to those who suppose (erroneously)
that what is now called "antique" furniture was
cheap when new. On the other hand, Mr. Luck, of
"the original carpet warehouse," Cheapside, provided
"a fine stout large Turkey carpet" for £11 11s. 0d.;
a sum which, regard being had to quality, compares
favourably with present prices.

Thomas White, who was suffering still from ague,
contracted probably when visiting his Essex property,
and his daughter visited her uncle in the summer
of this year. Writing to her brother at Fyfield, Miss
White remarks—

" We have had a great deal of visiting ever since our
arrival. . . . yesterday the family from Hartley drank tea
here. . . . I must mention now my uncle's new room, which
is quite finished in every respect and looks very handsome
indeed. The paper, a sort of light brown, with a coloured
border is extremely elegant, and the glass and other furniture
are very neat and handsome; in short the *tout en semble*
has a very pleasing effect, and it is I think one of the
pleasantest rooms I ever was in."

Writing on August 16th, 1780, Mulso remarks
upon the excellent flavour of the snipes he had
recently eaten at the Bishop's at Farnham, and
continues—

" I do not remember your ever shooting a snipe at Oxford
in summer, where there used to be plenty in winter; at that
time you used to practise with your gun in summer to
steady your hand for winter, and inhospitably fetch down
our visitants, the birds of passage: what you was then is
my son John now; I see him with his rod and line at the
Canal, and his gun lodged against a tree, a complicated
murderer."

To the Rev. R. Churton.

Fyfield, near Andover, Hants.

Aug. 31, 1780.

Dear Sir,—Your favour of July 10th carryed with it a
very obliging air, because it seemed to imply that you will
endeavour to pay me a visit.

Now let me (as old men love to be didactic) enjoin you to
leave the North as soon as you conveniently can, and to get
to Selborne by the last week in September at farthest;
for it seems to me to be very unreasonable to desire you to
come so far only for a week or a fortnight. About the

time that term begins I should be glad also to go to Oxford, and, provided health permits, will give you a cast in a post-chaise about the 12th or 13th of October all the way to College.

Dr. Chandler left me the week before last. After much delay we got one vol. of Bishop Beaufort's Register, the only one that can be found; but it contained only thirteen years of a long episcopate of above forty. It did not afford much concerning Selborne, but would, it seems, furnish much matter concerning the Lollards, who were cruelly harassed in the reign of Hen. 4th.

The way to Selborne is *Dorchester, Wallingford, Pangborn*; here leave the Reading-road, and go down the new turn-pike for *Aldermaston*-wharf, *Aldermaston*; *Basingstoke, Tunworth-down* under *Hackwood*-park pales, the *Golden-pot* ale-house, *Alton, Faringdon, Horse-and-Jockey, Selborne.*

Please to direct to me as before at Selborne near Alton Hants. If you know anybody in the N. whom it may concern, you may assure them that the crop of hops in the S. is prodigious; and that they are very fine in quality.

I conclude   Your most humble servant,

           Gil. White.

Pray write soon.

To Miss White
 At the Revd. Mr. White's
  At Fyfield near Andover.
 (*Turn at Harford Bridge.*) Selborne, Sep. 13, 1780.

Dear Molly,—My journey, I thank God, proved much less irksome than I had reason to fear: while I was in the carriage I was easy, but had a pinch at Winton, where we dined. My driver was the most civil man alive; but I found that the days were too short for one pair of horses; for, had there been no moon, we had been in *dead darkness*

all the way from Hartley,* where the day closed upon us.
I got to my house just at eight o'clock; and found my
nephew Barker, who arrived at ½ hour after six.

We are to make our farewell visit to Hartley in a day
or two, because the family are on the wing: they acknowledge *now* that they fear they shall have *two* houses on their
hands for four years; but Mr. Wilmot comforts himself
that he can come over and shoot pheasants at Hartley! Tell
your father that the state-prisoner lately released in Russia
is the father of the little Emperor Iwan; he and his family
were shut up in 1741, when Empress Elizabeth came to the
throne.

Triple Ladies' traces now blow in abundance in the Lythe.
Brother Ben. and family are expected next week; but we
see nor hear naught of nephew Ben.

Pray write to me *soon*, and send me the news of Fyfield;
and in particular let me know the state of your father's
health. Present my respects to my brother and sister H.,
and tell them I am indebted to them for all their good
offices; and in particular for the extraordinary trouble
occasioned by my indisposition.

> With all due respects I remain
> Yᵣˢ affectly,
> GIL. WHITE.

Timothy the tortoise was the subject of many experiments by his master, who at this time records—

"Sept. 17. When we call loudly through the speaking-
trumpet to Timothy he does not seem to regard the noise."

On September 21st Mulso wrote—

"Pray does your book come out this winter? I really
cannot hold out any longer. If you spoil the genuine

* Through the (now disused) "hollow lane."

elegance, and neat simplicity of the original design, by a farrago of antiquities, routed out of the rusts and trusts and crusts of time, I shall not esteem it so well as I once did, and so I tell you. Remember that Tom Warton has given the world two large specimens of his old bards and untunable harps.* Go to!"

To Miss White.   Selborne, Sept. 30, 1780.

Dear Molly,—Your letters are always agreeable to me: but your last was particularly so, because it brought so good an account of the state of your father's health.

Finding that Larby alone would never finish his job, I hired a whole band of myrmidons and set them to work on the *Bostal*,† where they have made great dispatch, and have but half a day's work to come, which has been delayed by the rains. They ran thro' the upper part a day sooner than I expected, because as we advanced, the soil grew shallower: but then we have been obliged to widen and raise *all* Larby's first attempts; because his part was so narrow, hollow, and clayey, that it soon grew dirty, and would have been impassable. By and with the advice of our Privy Council we took a higher direction than was at first marked out, because it much shortened the path, and brings us out straight at the top of the *slidder* before you come to *shop*-slidder at the corner of the Wadden. In our progress we found many pyrites in the

* The second volume of Thomas Warton's monumental work on ' The History of English Poetry' had appeared in 1778.

† The Bostal, described in the *Naturalist's Journal*, Sept. 27th, 1780, as "a sloping path up the Hanger from the foot of the Zigzag to the corner of the Wadden, in length 414 yards. A fine romantic path, shady and beautiful," is very well known to visitors to Selborne. It presents a much easier way of reaching the pretty common above the wood than the zigzag path, which was made when the writer of the above letter was a young man, and its construction now must be regarded as a sign of old age. The expense of making it was borne by Thomas White.

clay as round as a ball; and some large *Cornua Ammonis*
in the chalk. All people agree where party does not inter-
pose, that it is a noble walk: but there is a junto against it
called *Zigzaggians*, of which Mrs. Etty is the head; but
Mr. E. and Mr. Yalden would be *Bostalians*—if they dared.
The tall trees in the hanger are very fine when you are
among them, and the views through them romantic. My
Nephew Barker sets out for Fyfield on Monday, and regrets
that he shall miss of you and your father. Brother Ben.,
&c., came to Newton on Thursday. Pray desire your father
to receive my dividend at his leisure.

The bostal measures 400 yards, and the zigzag, which
is to be nicely cleaned out, 426.

Mrs. Etty and her young people set out to-morrow for
Oxford: Mr. E. is already in Oxfordshire.

All join in due respects, and good wishes.

Y^r loving Uncle,

Octbr. 1st. Gil. White.

Pray write soon.

My barometer is this evening at 28·6.

Thomas's brother has his ague still: he has taken the
roots of daffy-down-dillies.

To Samuel Barker.   Seleburne,* Novr. 23, 1780.

Dear Sir,—Your letter, though rather late, was very
acceptable. I was glad to hear that you had a safe and
pleasant journey back, and that you were so well pleased
with your journey into Hants, as to be able, on a retrospect,
to speak of it with some degree of satisfaction. The test
will be whether you liked your late reception by expressing
a willingness to come again. Pray give my respects to
Mr. Brodrick, and tell him that I always esteem my friend's

---

* The perusal of old documents, for the antiquities of Selborne, no doubt
accounts for the changed spelling now adopted.

friends; and therefore if he will come over next summer,
when you are here, from Pepperharrow for a night or two,
that I shall be glad to see him: and we will show him
some such prospects in these parts as may not be unworthy
his attention. To say the truth, the lower part of the
Bostal began to be dirty so that the Zigzagians (who
have horns and hoofs) began to triumph. Many of them,
in the shape of horses and heifers, ran up and down it,
doing it great damage with their feet: but to silence all
clamour I had all the bad part well-bedded with a quantity
of fern. Since this amendment Mrs. Etty and her sister
Stebbing, and Mrs. Yalden have been up and down it
by night and by day: so that party feuds are like to be
at an end. You do not, I hope, flatter me about my
Natural History: if you do not, I am much pleased to
find that an intelligent person like yourself approves of
it. Were it not for want of a good amanuensis, I think
I should make more progress: but much writing and
transcribing always hurts me. All I know about the sleep
of fishes is, that at the Black-Bear-inn in Reading there
is a stream in the garden which runs under the stables,
and so under the road into the meadows; it is a branch
of the Kennet. Now this water all the summer is full
of carps, which roll about, and are fed by travellers, who
divert themselves by tossing them crumbs of bread. When
the cold weather comes, these fishes withdraw under the
stables, and are invisible for months; during which period,
I conclude, they must sleep. Thus the inhabitants of the
water, as well as of the *air* and the *earth*, retire from the
severity of winter. Timothy, your friend, retreated into
his *hybernaculum* last week: he is laid up in the fruit-
border, in a dry, wholesome, sunny spot: at Ringmer he
was forced to lie in a swamp. My nephew Richard has
been here: he was quite transported beyond himself with
the pleasures of shooting; and, after walking more than

a hundred miles, killed *one woodcock*; which ill-fated bird
took the pains to migrate from Scandinavia to be slain
by a cockney, who never shot a bird before!!! Pleasure
is a most arbitrary matter! The pains my nephew took
in his new pursuit would have been a great misery to
many.   I conclude
                    Y^r affectionate friend,
                              GIL. WHITE.

I frequently see that new vegetable that we talked of
called a *Quid*, lying in a path: *habitat intra labra immundi
hominis*. I made some remarks of moment on the house-
martins just before they withdrew. They do not amount
to proof; but the presumptions are very strong indeed.

Now the leaf is down the Bostal discovers itself in a
faint, delicate line running up the hanger, such as would
require the hand of a Grimm to express it.

The *Naturalist's Journal* records the remarks on
the house-martins above referred to—

"1780, Oct. 13, 14. On these two days many house-
martins were feeding and flying along the hanger as usual,
'til a quarter past five in the afternoon, when they all
scudded away in great haste to the S.E. and darted down
among the low beechen leafy shrubs above the cottages at
the end of the hill. After making this observation I waited
'til it was quite dusk, but saw them no more; and returned
home well pleased with the incident, hoping that at this late
season it might lead to some useful discovery, and point out
their winter retreat. Since that, I have only seen two on
Oct. 22 in the morning. These circumstances put together
make it look very suspicious that this late flock at least
will not withdraw into warmer climes, but that they will
lie dormant within 300 yards of this village."

The incident was not forgotten, since he recorded—

1780   SEARCH FOR HOUSE-MARTINS   59

"1781, Ap. 5. Searched the S.E. end of the hanger for house-martins, but without any success, tho' many young men assisted. They examined the beechen-shrubs and holes in the steep hanger.

"[Ap.] 11. While two labourers were examining the shrubs and cavities at the S.E. end of the hanger, a house-martin came down the street and flew into a nest under Benham's eaves. This appearance is rather early for that bird. Quæ. whether it was disturbed by the two men on the hill?"

To the Rev. R. Churton.   Seleburne, near Alton, Hants.

Dec. 7, 1780.

Dear Sir,—If you have no more fears about a winter-journey than I had at your time of life, you might, I should hope, favour me with a visit during the approaching vacation. The country indeed is now shorn of its tresses, and much in dishabille; but we have still pleasant footpaths, wild views, and chearful neighbours. I will give you some roast-beef, plum-pudding, and other Xstmass-cheer. We do not, I believe, now keep the good season that is advancing so jollily as you do in the N.; but you will, I hope, be pleased with visiting Sir Adam de Gurdon's hall, where that old baron probably entertained his tenants with an ox roasted whole, and floods of brown ale. What I want is for you to try your hand at this place at this disadvantageous season; and then I shall not doubt but you will like it better in the summer. We have finished a walk of 400 yards in length through an hanging wood just above my house; which we are apt to think will please strangers, because we like it ourselves. From hence we look on the village in a very pleasing light. If you are a draughtsman, I can show you some stained views taken from nature by an artist that came down to me from London.

My progress in Nat. Hist. is very slow indeed. I now

and * . . . . . . advertised, I see, and will be out
in February. I heartily wish he may give no reason
for complaint with respect to religious matters: in other
respects he will be secure of fame.

If I was to meet Gen. Arnold† I should address him
thus:—

> "But wherefore thou alone? wherefore with thee
> Came not all?   *   *   *   *
> *   *   *   'had'st thou alledg'd
> To thy deserted host this cause of flight
> Thou surely had'st not come sole fugitive."

I am, with due respect,
Your most humble servant,
GIL. WHITE.

Mr. G.[ibbon], I understand, will draw a comparison between Xstianity papal, and Mohammedism; and indeed I
am at a loss to say which will make the most hideous
picture. I mean the popery of the darker ages.

To the Rev. R. Churton.   Selborne, Dec. 19, 1780.

Dear Sir,—By your letter of the 14th to Dr Chandler,
which the D^r has communicated to me, I am glad to find
that you are so well disposed to make me a visit, and hope
you will meet with no interruption. You will not, I hope,
over-stay this unprecedented run of fine weather, that has
befallen us now for more than three weeks, without rain,
wind, or frost!

If you have a friend in London to whom you can send
your portmanteau, then you need only desire him to direct
it for you "at the Revd. Mr. W. at Selborne, to be left at
the Swan-Inn at Alton, by the Southampton coach," which

* Letter imperfect. The missing words, no doubt, referred to the second
and third volumes of Gibbon's History, which was published in April, 1781.

† Benedict Arnold, an American general, in 1780 fled precipitately to the
British lines, after his plot to betray his countrymen at West Point had
been discovered.

comes from the Belle Savage-Inn on Ludgate hill; but if you have no such person, then direct it to Mr. Edmund White, at Mr. Hounsom's mercer in Fleet-street, London, to be forwarded to Mr. White &c., by the Southampton coach.

If you call at Caversham, pray present my most respectful compliments to Mr. Loveday, and the ladies. I have not the pleasure to be known to D[r] Loveday.

Your most humble servant,
GIL. WHITE.

MRS. SNOOKE'S HOUSE AT RINGMER IN 1783.

CHAPTER III.

THE following letter, signed "V." (= Vitus), the
original of which, Mr. Moy Thomas, in his memoir
of Collins, mentions having seen in Gilbert White's
handwriting, appeared in the 'Gentleman's Magazine' for 1781 (vol. li. pp. 11, 12). It is interesting
as showing his acquaintance with William Collins,
who was a Chichester man, but not quite accurate in
all details concerning the career of the unfortunate
poet, who died in 1759. What called forth the
letter does not appear, but Langhorne published an
edition of Collins' poems in 1781, and Johnson's
life of Collins (in his 'Lives of the Poets')
appeared in the same year :—

January 20th.

Mr. Urban,—William Collins, the poet, I was intimately
acquainted with from the time that he came to reside at
Oxford. He was the son of a tradesman at the city of
Chichester, I think an hatter; and being sent very young to
Winchester School, was soon distinguished for his early
proficiency, and his turn for elegant composition. About
the year 1740 he came off from that Seminary *first* upon

the roll,* and was entered a commoner of Queen's College. There, no vacancy offering for New College, he remained a year or two, and then was chosen a demy of Magdalen College, where, I think, he took a degree. As he brought with him (for so the whole turn of his conversation dis- covered) too high an opinion of his school acquisitions, and a sovereign contempt for all academic studies and discipline, he never looked with any complacency on his situation in the University, but was always complaining of the dulness of a College life. In short, he threw up his demyship, and, going to London, commenced a man of the town, spending his time in all the dissipation of Ranelagh, Vauxhall, and the playhouses; and was romantic enough to suppose that his superior abilities would draw the attention of the great world, by means of which he was to make his fortune.

In this pleasurable way of life he soon wasted his little property and a considerable legacy left him by a maternal uncle, a colonel in the army to whom the nephew made a visit in Flanders during the war. While on this tour he wrote several entertaining letters to his Oxford friends, some of which I saw. In London I met him often, and remember he lodged in a little house with a Miss Bundy, at the corner of King's Square-court, Soho, now a ware- house, for a long time together. When poverty overtook him, poor man, he had too much sensibility of temper to bear with his misfortunes, and fell into a most deplorable state of mind. How he got down to Oxford I do not know; but I myself saw him under Merton wall, in a very affecting situation, struggling, and conveyed by force in the arms of two or three men, towards the parish of St Clement, in which was a house that took in such unhappy objects; and I always understood that, not long after he

---

* Mr. Joseph Warton, now Dr. Warton, headmaster at Winton School, was at the same time *second* upon roll; and Mr. Mulso, now (1781) pre- bendary of the Church at Winton, *third* upon the roll.—V.

died in confinement, but when, or where, he was buried I never knew.

Thus was lost to the world this unfortunate person, in the prime of life, without availing himself of fine abilities, which, properly inspired, must have raised him to the top of any profession, and have rendered him a blessing to his friends, and an ornament to his country. Without books, or steadiness or resolution to consult them if he had been possessed of any, he was always planning schemes for elaborate publications, which were carried no further than the drawing up proposals for subscriptions, some of which were published; and in particular, as far as I remember, one for a 'History of the Darker Ages.'

He was passionately fond of music; good natured and affable; warm in his friendships, and visionary in his pursuits, and as long as I knew him, very temperate in his eating and drinking. He was of moderate stature, with grey eyes, so very weak at times, as hardly to bear a candle in the room; and often raising within him apprehension of blindness.

With an anecdote, respecting him, while he was at Magdalen College, I shall close my letter. It happened one afternoon at a tea visit, that several intelligent friends were assembled at his rooms to enjoy each other's conversation, when in comes a member* of a certain College, as remarkable at that time for his brutal disposition as for his good scholarship; who, though he met with a circle of the most peaceable people in the world, was determined to quarrel; and, though no man said a word, lifted up his foot and kicked the tea table and all its contents, to the other side of the room. Our poet, though of a warm temper, was so confounded at the unexpected downfall, and so astonished at the unmerited insult, that he took no notice of the aggressor, but getting up from his chair calmly, he

* [Hampton.] The translator of Polybius.—V.

began picking up the slices of bread and butter, and the fragments of the china, repeating very mildly—

"Invenias etiam disjecti membra poetæ."

I am your very humble servant,

V.

The following entry in the *Naturalist's Journal* may interest Selborne residents :—

"Jan. 26 [1781]. My *Heliotrope*, which is J. Carpenter's workshop, shows plainly that the days are lengthened considerably : for on the shortest day the shades of my two old chimneys fall exactly in the middle of the great window of that edifice at half an hour after two p.m., but now they are shifted into the quick-set hedge, many yards to the S.E."

To Miss White. Selborne, Feb. 6, 1781.

Dear Molly,—I was much pleased to see a letter from your father under his own hand : for it was indeed a very long time since I had seen any such thing. Pray desire him soon to receive my Xmass dividend, because I want to get all my monies together that I may pay my debts. We have had a dry and a mild winter : from Nov. 25th, 1780, to Jan. 18, 1781, there fell, with us, only ·69 of rain ; and snow ; less than three-quarters of an inch !

We were much pleased with your account of *Buddle* : Dr Chandler calls you *Sister* Antiquary, and talked much of you this day : he wonders your father does not get acquainted with Dr Ducarrel,* whom he esteems as a very knowing man, and one from whom much might be learned.

Yesterday Miss Shutter was married at Beaconsfield to a Mr. Ransford, a young gentleman of a very great landed estate near Northampton : his mother lives at Bath. Yesterday also Mr. Etty received a letter from his son

* Librarian at Lambeth Palace.

Charles, dated in July last, at sea within two degrees of the line. It came from Cork, and expressed that, so far, they had experienced a pleasant voyage; and were not to stop 'til they arrived among the Indian islands.

Pray, niece, write to me. The boys might prepare their skates, but they would be troubled to find water to make ice this winter. Our ponds were dry, and the millers wanted water for grinding.
Y^r loving uncle,
GIL. WHITE.

On November 21st in the previous year (1780) John White, who had long been a sufferer from a serious rheumatic complaint, died at Blackburn. He was buried under the Communion table of the parish church there. A mural tablet records that he was " an ingenious and accurate Naturalist."

Mulso writes on February 11th, 1781 :—

" As his constitution was irrecoverably injured, his release was a blessing to himself, as a very worthy man. But his family and friends miss him much; and I think the world has a loss in him, for he was a man of more than private accomplishments, and united in himself things which do not commonly assemble, mathematics and poetry, philosophy and humour. Pray what is to become of his 'Fauna'?

"That work is not, I hope, to be secreted, like a certain person's, whose false modesty will not trust forth a Piece really good for fear it should not be absolutely perfect, which would be *prodigii instar*."

The 'Fauna Calpensis' was never published. In a notice of John White, written in a pedigree compiled, apparently in 1826, by his nephew John (son of Benjamin) White, who, in partnership with his brother Benjamin (who died in 1821), carried on their father's publishing business, it is stated that the work was "now existing in MS."; but it seems since then to have been certainly lost or destroyed, very possibly in 1839, when the last member of the White family ceased to reside in Gilbert White's house at Selborne, where, there is reason to believe, the MS. had remained. The introductory chapter describing the Rock of Gibraltar, in John White's handwriting, still exists.

In the course of the year following her husband's death Mrs. John White came to Selborne, where she resided with her brother-in-law during the rest of his life.

From the *Naturalist's Journal* :—

"Feb. 10. The nuthatch brings his nuts almost every day to the alcove, and fixing them in one corner of the pediment, drills holes in their sides, and after he has picked out the kernels, throws the shells to the ground."

Writing to his nephew, Samuel Barker, who had inquired whether anything corresponding to the aurora borealis was ever seen in the southern hemisphere, Gilbert White quotes a passage from J. R. Forster's 'Observations in a Voyage round the World,' p. 120, and then proceeds as follows :—

To Samuel Barker.   S. Lambeth, Mar. 26, 1781.

My thanks are due for your entertaining account of the *Testudo aquarum dulcium.* You do very right, I think, in looking into history, which is a very gentleman-like study.

You, who have youth, health, and a strong retentive memory on your side, will soon make a vast progress.

Pray tell your mother that I thank her for her letter. Jack White's time will not be out 'til the 16th of next June, when he and his mother will come Southward among their relations. What mode of life that young man will take up I have not yet heard; whether he will walk the hospitals in town, or become for a time a journey-man. Poor Joe Woods, son of Mr. Jos. Woods, a promising young man of 21, is just dead of a decline, to the great sorrow of his parents, &c.

>With all due respects I remain
>
>Your affect. friend,
>
>GIL. WHITE.

I propose to return home on Thursday.

Having had no rain, not once enough to measure, at this place, since the last week in Feb., the degree of dustiness is horrible, and not to be described. As brother Thomas and I walked out this morning a gale rose from the N., which filled the whole atmosphere with such a cloud from road to road that the prospect was quite obscured!

On the 27th of Feb., Tuesday, the day I left Seleburne, we had such a terrible storm of wind that vast mischief was done in the S. and W. of England. I expected to hear of great damage, especially in Sussex; but was thankful to find that I had escaped with the overturning of my alcove into the hedge, the overthrow of my stone-dial, and what grieves me most, because it cannot be repaired, the ravage of my great wal-nut-tree, which, they write word, is almost torn to pieces! The gale began at eleven a.m., the wind W.; but the great damage was done about five p.m., the wind N.W. Soon after a calm succeeded. Derham remarks that most tempests from the W. vere a little at last to the N.W., and then the ravage and damage takes place.

Pray write, and on *large* paper.

To Miss White.   Selborne, April 9, 1781.

Dear Molly,—It is full time that I should sit down, and return my best thanks to your father and you, and my other relations for the many good offices that I experienced at S. Lambeth and London.

On the road I found the dust very troublesome: yet I got safe to Alton at five o'clock. You need not for the future suffer any inconvenience on similar occasions, for Mr. Edwards, the Hants Map-man, is going to make a canal from Chertsey to Alton; so that you may come all the way as far as Alton by water. Moreover, he intends to bring a canal up to Bishop's Sutton, Bassat's village, from Winton; and then to bore a hole for five miles from Alton under Bentworth, Medsted, &c., and out at Ropley-dean; and so to make a junction of the Wey, and the Itching. You may smile at this proposal; but the projector tells me that the tunnelling part will be the most profitable and easy to be managed; because the chalk, when burnt into lime, and the ashes of 4,000 barges of peat made by the burning, will, at half the present price of lime and ashes, produce £31,371; besides infinite advantage to the lands! Edwards has published a pamphlet on this subject, to which I refer you; and is to have a county-meeting soon.

But there is a matter at present of more consequence to me, concerning which I wish you would consult your father, and write me word by *the first post*; and that is that I sent Goody Hampton, my weeding woman, last Wednesday to the post office with a letter to your uncle Harry, in which I enclosed, or thought I had enclosed, *half* a £10 bank-note. But lo, nephew Samson* is come up express to tell me that the letter came safe and unrumpled, but in it no half bank note. Since my nephew came I have examined every probable place, but to no purpose. The

---

* Son of Henry White of Fyfield.

only solution of this difficulty seems to be, if it so prove, that at the same time that I made up the Fyfield letter, I made up also a frank to Mrs. Bentham,* in which were two or three little papers. Now it is possible the half bill may be gone to Oxford. If it should prove so, it will also prove that my memory is very bad. But if this half note is not forthcoming, let me know what I am to do with the remaining part, whether it should be sent to town. The end containing the number and person to whom drawn to is missing. I cannot suspect Goody Hampton: her honesty and ignorance acquit her. Brother Harry's servant took the letter at Mullen's pond of the post boy, so no fraud could be committed at the turnpike house.

Sam. has brought me up the dog Rover, whom, if he behaves quietly, I shall approve of, because he is both large, and good for nothing; I mean has no sporting blood in him.

Y^{rs} affect.,

GIL. WHITE.

To the Rev. R. Churton.

Selborne, May 9, 1781.

Dear Sir,—When I called at Brazennose College in the Easter week, I was sorry but not disappointed in not finding you, because Mr. Loveday had intimated that probably you would be gone on a visit to his son.

As you have seen Selborne, and the nakedness of the land at Xmas, you will not do it justice if you do not come and visit it in all its glory, in its full foliage, and verdure.

I therefore exhort you and enjoin you to come and spend the Whitsun vacation here, where your company and conversation will be very acceptable; and, if I mistake not, my neighbours will be glad to see you also.

---

* Wife of Dr. Bentham, Gilbert White's tutor when at Oriel.

If you come by Caversham, be pleased to ask for a parcel of papers which I left with Mr. Loveday.

> I am, with due respect,
>> Your most affectionate servant,
>>> GIL. WHITE.

If you will direct your portmanteau to be left at the Bell Savage on Ludgate hill London, to be forwarded to the Swan at Alton by the Southampton coach, it will, I trust, come safe.

On June 16th, 1781, Mulso wrote to thank his old friend for his interposition with Dr. Sheffield, now Provost of Worcester College, in favour of his son, John Mulso, junior, who had been elected to a scholarship there—

"I presume this will find you at Selbourne after your visits in London and Surry. . . . Pray give me an account of your family and their proceedings, and how Jack Gib. goes on. I daresay well, and hope he will be a comfort to his mother.

You have robbed the good old Bishop* of a pleasure by deferring the publication of your book. Are you cowardly, or are you over nice and curious? Make haste, my dear old friend, or you may rob the nephew too. Am I not three score in Novr next? Do you keep it for my chair-days? Perhaps you mean to assist my ideas, when I cannot expatiate to enlarge my observations. I do not know that I could conquer Selbourne Hanger now."

On July 28th the *Naturalist's Journal* has a curious note on the sense of colour in birds :—

"The white throats are bold thieves; nor are the red

---

* John Mulso's uncle (both in blood and by marriage), Dr. Thomas, Bishop of Winchester, who had recently died.

breasts at all honest with respect to currans. Birds are guided by colour, and do not touch any white fruits 'til they have cleared all the red; they eat the red grapes, rasps, currans, and goose berries first."

To Miss White.   Selborne, Aug. 1, 1781.

Dear Niece,—I thank you for your care about my tea and chest, which Thomas is to fetch from Alton this afternoon. We shall be very glad to see you and your father whenever it is convenient, and hope we shall meet happily together. Mr. and Mrs. and Miss Fort, and Mr.* and Mrs. and Miss Hounsom, &c. dined with me last Saturday; and this day they set out for Funtington, from whence Mrs. and Miss Fort are to return to Newton.

We are in the midst of wheat harvest, and have glorious weather: the wheat in general has much straw; but proves light and blasted; but in some countries the state of the crop is deplorable. Our gardens are burnt up; but John† says that we are very verdurous in comparison with South Lambeth.

Tell your father that I thank him much for his prefaces,‡ some of which have given me great pleasure. He need not, unless he chuses it, order a newspaper down; for we have four every week. I am sorry for Uncle Harry's disappointment: I thought the young man's coming was a settled thing. Mr. Etty is in Oxfordshire, whither he was called by the sudden death of his tenant, who was not in good circumstances.

Mrs. Hounsom came down to Newton in the coach, in

* Gilbert White's niece, Anne Woods, married John Hounsom of Funtington, Sussex, in 1792. His father, the "Mr. Hounsom" of this letter, was probably the "Mr. John Hounsom, linen draper in Fleet Street," mentioned in Gilbert White's letter to his brother John of March 9th, 1775.

† "Gibraltar Jack."

‡ T. Holt-White mentions "my father's [*i.e.* Thomas White's] Preface to his republication of Evelyn's 'Fumifugium.'" This edition is not known at the British Museum.

which was a female fellow-traveller, who after a time told her she was going to Selborne in Hants. Pray, says Mrs. H. do you know Mr. White of that place? Yes, replies the woman, by character; that is the gentleman that *starved his niece*. Sure, replyed Mrs. H. you must be mistaken; I can hardly credit the report. You may depend on the truth of it, rejoined the person again, for I have relations at Selborne, and go there every year. Thus you see, that you have been looked upon as one of the *Children in the wood*, and I as the *Unnatural Uncle*. I must desire you therefore to come down as plump and chearful as possible; and to eat and drink plentifully all the time you stay, that I may no longer labour under the atrocious imputation of starving my relations!

My Sister, and John came here yesterday to dinner, and join in respects to you and your father, and nephew and niece Barkers.　　　　　　Yʳ hard-hearted Uncle,

GIL. WHITE.

Miss Etty is at Priestlands bathing.

I rejoice to hear that Miss Mary Barker's hand is so finely healed.

From the *Naturalist's Journal*—

"Aug. 24. One swift still frequents the eaves of the Church; and moreover has, I discover, two *young* nearly fledged which show their white chins at the mouth of the crevice. This incident of so late a brood of swifts is an exception to the whole of my observations ever since I bestowed any attention on that species of *Hirundines*!"

To Miss White.　　　　　　Selborne, Sep. 4, 1781.

Dear Molly,—For some weeks past I have now expected a letter from you every post; and for several evenings past have flattered myself that you and your father would surprize us agreeably by coming without any notice. Mrs. Etty, who is a wise woman of Tekoa, told me positively that you would appear last Saturday. The great heats are now abated, and the dust is layed; but the lovely weather continues. We all therefore earnestly wish to see you both, while the country is so agreeable. Peaches and nectarines which are delicate, are going out; but grapes will soon come; they look black. We have caught 20 hornets; wasps there are none.

Pray enquire of your father if he has received my dividend; because I want to pay John Stevens. Tell him I thank him for ensuring Iping buildings; and enquire if he has been so kind as to pay the continuance of the Selborne, and Harting ensurance, which was forgotten.

Your father's hazel-stick is looked out, and wiped; and the Bostal is in good order.

With all due respects I remain

Y^r loving Un*k*le,

(for that is the modish way of spelling the

relation-ship)

GIL. WHITE.

Mrs. Etty is very angry!

A letter to you, post mark Gloucester, is lately arrived.

On September 29th, 1781, Mulso writes:—

"I thank you, in his name and my own, for your late civilities to my son, of which he is very full; I think he has even brought home with him the tone of your voices, your phrases, and your stories. He is likewise sensible of the charms of Miss White, and the obligingness of your neighbours. He delivered your Piece by Mr. Grimm (the Temple*) which I approve of very much; though I still think that Mr. Grimm has a heavy hand at a *distant* view; nor can I forgive him, but as a Christian, for giving so

---

* This became Plate VIII. in 'The Natural History and Antiquities of Selborne,' p. 343.

little an idea of the *high point* of your Hermitage.* In the place he is just, but gives no representation of the position with regard to the lower grounds. In the Temple, by shewing the turn of the Hangers, and by multiplying the grounds before you, he describes the advanced ground that you are upon. Colouring would express it compleatly, but the engraving is of too uniform a shade to do it justice.

" I was exceedingly obliged, my dear old friend, by your visit to me; especially considering that I have seemingly been negligent with regard to visiting Selborne."

During the visit referred to in the next letter the *Naturalist's Journal* records the opening by " brother Thomas " of two tumuli on Selborne Down ; "nothing found."

To Miss White.   Selborne, Nov. 13, 1781.

Dear Molly,—Your visit, which you call a long one, I call a very moderate one, and wish you could have been prevailed on to have stayed longer: however I thank you and your father for coming to see us. Mrs. Yalden lately took a handful of sticks, and stuck them along the common down to the Bostal, and from the mossy-dells to the zig-zag; she afterwards took a cartful of chalk and a carter, and ordered him to lay lumps of chalk all the way, for direction posts, the whole length of the down, so that Mr. Etty who used to say he would not go over the common by himself in the dark for £50, might now venture for half the money.

Molly Berriman† continues to be the most unfortunate of women, for now she has lost *all* her clothes. When the

---

* This sentence locates the site of the Hermitage, *i.e.* the *original* Hermitage, well known to Mulso. The curious visitor to Selborne will readily distinguish the "area" (as Gilbert White called it) on which it stood, cut out of the chalk hill high up, a little to the west of the Zigzag.

† The wife of a Selborne farmer.

soldiers left this place, two maidens of the village followed them; and that they might cut a figure in their new way of life, stripped the poor woman's wardrobe. Rob. B. pursued them in great wrath; and overtaking them at Hungerford, brought the damsels back. But as nothing was found upon them, after much trouble and expence he was forced to let the matter rest.

Poor Dame Larby gets worse; and must soon, it is feared, be starved.

We have had fine rains this month. On the 2nd 56; on the 5th ·78; on the 6th 1·21; on the 10th ·18; on the 11th ·58.

Y^r loving unkle, with a K,

GIL. WHITE.

Oct. 30. The tortoise went under the ground in his coop, but not liking his quarters, on Nov. 8 he lifted up his coop, and came forth, and has buried himself again in the laurel-hedge, where he will probably be lost in the profoundest slumbers during the uncomfortable months of winter.

Pray write to us.

Our grapes are good still, and not quite gone. We have eat of them twice every day to this time.

On November 21st, 1781, the Provost of Oriel College, Dr. John Clarke, died, and Dr. John Eveleigh was elected in his place on December 5th. It does not appear that Gilbert White, though he went to Oxford for the election, was a candidate on this occasion. Writing at this time to her brother, Miss White remarks—

"We see in the paper an account of the death of Dr Clarke and wish we could elect my uncle Provost of Oriel, but I fancy he would make some objections, was it in our power."

Probably he thought himself too old at sixty-one to undertake new and important duties, and had become too much attached to Selborne to leave it. That he was engaged in compiling his book is shown by the following letter of December 4th, 1781, from the niece mentioned :—

"Agreeable to your request I have written out the passages in Verstegan and Chaucer. . . . My uncle Benjn says he has seen the *grey* crow on Selborne-Common : we are in doubt whether it is in your list of birds.

"Pray add to your provincial words *Merise*, a small bitter Cherry says the French dictionary (perhaps from *Amarus*). . . . 'October had the name of Wyn-monat, and albeit they had not antiently wines made in Germany, yet in this season had they them from divers countries adjoining ; rather because in this month'—

'. . . pocula læti
Fermento atque acidis imitantur vitea sorbis.'

"I wish for Fermento we might read Frumento. . . .

"Extract from Chaucer's 'Floure and the Leafe.'

'And to a pleasant grove I gan to pas,
Long or the bright Sonne uprisn was ;
In which were okes grete, streight as a line,
Under the which the grasse so fresh of hewe,
Was newly sprong ; and an eight foot or nine
Every tree wel fro his fellow grew.
With braunches brode, laden with leves newe,
That sprongen out ayen the Sonne shene,
Some very redde, and some a glad light grene.'"

To Miss White.   Seleburne, Decr. 19, 1781.
Dear Mrs. Mary,

The young Antiquary,—Your letter of the fourth of this month afforded us much pleasure and information. Dr Chandler thinks that you now deserve more than ever to be made S. A. S. : by the first S. I suppose he means *Soror*.

As the Saxons had invented *significant* names for eleven of their months, I wonder that April should come off so poorly; for certainly the Goddess *Goster* is as *arbitrary* an appellative as June, July, August, &c.

I agree with you in preferring the reading of *Frumento* for *Fermento*; if any MSS. would keep us in countenance; or perhaps *fermento* in its place might be read *adjectively*, pro *fermentatis*. Virgil often expresses himself in that manner.

The poets have many times taken notice of the various shades and tints of autumnal leaves; but Chaucer, in your quotation, is the only one that I have remarked that has observed the different colours of leaves at their first coming out in the spring.

You are certainly right respecting *Merise*, a bitter cherry: hence no doubt comes our provincial word *Mery*, or more probably *Meris*, the S. being dropped.

Thanks are due for our salt-fish which came last Saturday, and looks finely; pray write word what it cost.

The girl has knit your father one pair of stockings, and almost an other. I will procure more worsted. Her eldest sister is hired as a nurse-maid to Mrs. Clement.* Poor Dame Larby lies still in the same hopeless way.

I have, in my time, seen two *grey* crows in Selborne parish. My well, and others, particularly that most profound one at *Heards*, continue very low. The stream at Gracious street just runs, and Well-head is not much increased. Mrs. White and Mr. and Mrs. Etty, who have been gossiping all the morning with Mrs. Clement, bring some imperfect accounts of good news both from India, and off Brest. God grant that they may be true! We often exceed you in rain. In Dec. .79 we had 6·28 in., and in Nov. .81 6·18 in. In this current month we have caught, as yet, only ·100. I have received from Shields the nursery man four peaches

* His niece, Jane, daughter of Benjamin White, who now lived at Alton.

and nectarines, trained trees, that are to bear next year: they have fine regular heads, but are very dear!

In a note to my account of Wolmer-forest* I have mentioned that some old people have assured me, that of a winter's morning they have discovered sunk trees in the bog by the hoar-frost that lay longer over the space where they were concealed, than on the surrounding morass. Nor does this seem to be a fanciful notion; but conformable to true philosophy. For Dr Hales says, "that snow lies longest, as he has often observed, over drains, elm-pipes, &c.: because these intervening, detached substances, interrupt the warm vapor from the earth, and impede the thawing." See Hales's 'Hæmastatics,' p. 360. Hence I make the following quære. Might not observations of this nature be extended to domestic uses by the discovery of old obliterated dreins and wells about houses; and in Roman stations, and camps, lead to the finding of pavements, baths,† graves, and other hidden relicks of curious antiquity? Pray continue your communications, and particularly respecting the quære.

Yr loving unkle,
GIL. WHITE.

To Miss White.

Selborne, Dec. 31, 1781.

Dear Molly,—The girl has finished only one pair of stockings, which I send; another pair is almost completed; when she has done what she is about, I will order more worsted. As soon as you know, pray send me the price of the salt fish.

Many thanks for your last letter. We should be glad of more hints, quotations, and anecdotes.

---

* Letter VI. to Pennant, 'The Natural History of Selborne.'

† Until well into the nineteenth century the remains of rooms of Roman houses in this country, which frequently contain hypocausts, were called "baths," since in the milder climate of Italy hypocausts rarely appear except as the foundations of baths. Hence the mistake.

As Uncle Will was coming down in great glee from Newton to Selborne, he attempted to get down the slider immediately below the hermitage. The slippery ground hurried him on much faster than he was aware, so that he fell with his head on a stump and cut his forehead sadly. He was stunned with the fall, and lay senseless for some time. I was shaving when he came in, and seeing at my elbow a courier with his hand full of letters, and all over dirt and blood, could not imagine at first who he was.

Y^r loving unkle,

GIL. WHITE.

We are glad to hear that your father is so well.

To Miss White.

Selborne, Jan. 16, 1782.

Dear Niece,—Dr Chandler and Mr. Churton went lately to visit Binstead * church, where in a room to the N. of the church, called the school-room, they found the following inscription in Saxon characters: "Richard: de: Westcote: gist: ici: Deu: de: sa: Alme: eit: merci! Amen." And on the tomb the stone effigy of a Knight Templar in his shirt of mail, with his legs across, and a little lion between them; with his shield on his arm, and his sword by his side; and all in pretty good preservation. A quarter of a mile to the W. of the church stands an old mansion called *Westcote*, perhaps once the residence of this Knight.

Well-head runs now very strong, and sends out perhaps 30 gallons of water in a minute; the stream nearly fills the cart-way.

The bostal, from much use, is dirty; while the zigzag, from the contrary reason, is sound and clean.

Dame Larby dyed yesterday morning.

---

* A village near Selborne.

Pray write me word what the salt fish came to.

After the fast Mrs. White and I hope to visit you.

Your affectionate Uncle,

GIL. WHITE.

The wells are now very high. We have some hopes of seeing Uncle Harry next week.

To Miss White.   Seleburne, Feb. 9, 1782.

Dear Molly,—That you may not be kept in any suspense, I think proper to inform you that Mrs. White and I propose to dine and *lie* at Alton on Monday the 18th. I believe I should have said *sleep*; but I cannot always promise to fulfil my engagements in that matter just to a night. However, the day after, we propose to reach S. Lambeth by tea time; where, I hope, we shall all have a happy meeting.

The Grange farm is now going to be sold in earnest: the poor owner has but seven years remaining of her lease of twenty-one; so that the purchaser must first apply to the Coll. and know their terms, and then treat with the tenant Mrs. W.

The other day I fetched up a bottle of brandy (for you know I deal much in brandy) when lo the contents, though all the rest had been bright, was of a deep purple! Why so, niece?

Venus was so resplendent last night, and is again this evening, that she casts a beautiful pale light on the walls of my chamber, &c., and shows distinctly the shades of the window-frames, and the lead between the panes of glass; or, to speak as an astronomer, she *shadows* strongly. Mars, who made such a figure in June, now looks very simple.

Mr. and Mrs. Clement left us this morning. We have now sharp frost! Your father, I hope, will receive my Xmass dividend.   Yʳ affectionate Unkle,

GIL. WHITE.

Writing from South Lambeth to her brother at Fyfield on March 11th, 1782, Miss White tells him that—

"My uncle and Mrs. John White intend returning next Friday. My uncle is obliged to leave us on account of his church: we have just heard of the death of Mr. Roman: he christened me. My uncle may perhaps lose his curacy of Faringdon. The living is, I believe, intended for a young man about seventeen."

Returning from Lambeth Gilbert White received a visit from his brother Henry, with whom, as the latter records in his own Journal on April 6th, 1782, he visited Winchester on his way to Fyfield, where—

"H. W. and G. W. saw the new altar piece at Winchester Cathedral, the raising of Lazarus by Mr. West—very fine, the frame gone to be changed."

On April 5th, 1782, Mulso writes :—

"I lately learned an event that I think must have interested you a good deal, and that was the death of Dr Roman; by which your curacy of Faringdon must have been hazarded. I have not yet learned who succeeds, or upon what footing you now are. I think your College might make an exchange for you, if the value of the living would compensate the Fellowship and curacy."

These remarks serve to show that the writer's scruples about the retention by his friend of his Fellowship must have been satisfied, as in truth they should have been.

On June 2nd, 1782, Mulso writes thanking his friend for kindly receiving the recommendation of a

Worcester College candidate for an Oriel Fellowship, in whom he was interested from the fact that the election of this young man at Oriel would facilitate the election of his son John to a Fellowship at Worcester College. He continues in the old strain—

"Another winter is passed without your *Essays*. I have no more to say than that you are a timorous provoking man. You defraud yourself of a great credit in the world : as to your labouring at your Antiquities, it is *mal-a-propos* ; the world does not care for such rough work now. Your Porch will be bigger than your House, and you will clap a Gothic Front upon a plan of Palladio.* I mean this, if you labour *too much* at it. I will give you credit myself that everything that comes from you shall be good. I shall not be quite sorry when you have left Faringdon, but I wish you a Sinecure in its room, if such a thing would not vacate your Fellowship. But perhaps you are like an old prisoner of the Bastille, and would fear to catch cold in your leg, if it had not a chain on."

To Miss White.   Aug. 3, 1782.

Dear Molly,—Pray desire your father to receive my midsummer dividend, and to bring me down *thirty pounds* in Cash.

Mrs. White desires you to bring her down one quire of *black-edged paper*.

If you have some *mushroom-spawn* to spare, I should be glad of some.

---

* The result shows that this prophecy of Mulso's, like others that he made of his friend's work, was correct. Very few people read the 'Antiquities' in comparison with those who read the 'Natural History' of Selborne. Yet they are good of their kind, and creditable to their author, as indicating the spirit of thoroughness with which he set about describing his native parish.

Thomas desires a parcel of *sweet-williams.*

Your father gave me two yellow asphodels; one of which has blown finely.

You may bring us half an hundred of good *loaf-cabbages,* and some *ferruginous fox-gloves,* if any to spare.

We shall hope to see you soon.

Mr. and Mrs. Richardson are here, and three young Wykehamists, for whose sake Timothy is locked safe in the stable. D^r and Mrs. Stebbing come on Monday. Mr. and Mrs. Powlett dined here yesterday.

We shall hope to see you and your father very soon.

Y^r loving friend,

GIL. WHITE.

The quantity of water that fell at Selborne in July last was 7·09 !!!   Multum computruit fœnum !

On August 7th, 1782, Mulso writes from Meonstoke :—

"You must be in great beauty in your verdurous spot. How are you in your health? and how able are you as a horseman? Do you amble about your neighbourhood? Do you yet serve Faringdon, or is the new representative of Mr. Cage come to exclude you, and to enlarge you? . . . You and your place are among the prejudices of my youth, and my mind dwells upon them with a fondness that I do not feel for newer and grander things.

"Am I to die before your little favourite work comes out? *Des aliquid famœ* and don't be so tedious and phlegmatic."

During the summer, which was exceptionally wet and cold, several relations visited Selborne. The *Naturalist's Journal* records—

"Sept. 1–7. The swifts left Lyndon in the County of Rutland, for the most part, about Aug. 23. Some continued till Aug. 29: and one till Sept. 3rd!! In all our observation Mr. Barker and I never saw or heard of a Swift in Septem^r. tho' we have remarked them for more than 40 years.

"Sep. 15. My brother Thomas White opened two of the most promising tumuli on the down above my house" [without result, however].

On October 22nd, while staying with his brother at Fyfield, he made the observations on goldfish, which appeared in 'The Natural History of Selborne,' Letter LIV. to Barrington.

Everybody in the Eastern Counties seems to have his own particular cure for ague, and it would seem that Thomas White, who had suffered sharply from this complaint, formed no exception to the rule.

To Miss White. Selborne, Dec. 7, 1782.

Dear Niece,—Your father will be pleased, no doubt, to hear that his good offices to Anne Osgood were effectual. This woman went to Burbey,* and took one ounce of the red bark; but still the ague returned; though with less violence. She then entered upon and took all the second ounce which performed a compleat cure. As a corroborating circumstance Burbey gave her a warm linsey-woolsey jacket: before she had no gown to her back. Your father has often enquired of me whether the autumn previous to the severe winter 1739–40 was wet or not: all that I could tell him was, that the lavants at Chilton Candover were high, when the rigorous season began. But now I see by

---

* Who kept the village shop at Selborne.

Dr Huxham,* that, at Plymouth at least, the whole year 1739 was a wet one: for the rain in that period was 36·308 in., a large quantity. I have been much entertained with the remarks of that accurate writer; and find from him that the district of Plymouth is rather to be called a wet one from the frequent rains, more than from the quantity. Besides, I see, in our very dry years they had little rain: as in 1741, 20·354 in.; and in 1743, 20·908 in. Nor do I find that in the 20 years and upwards that the D^r carried on his exact measurings, that ever the water was caught that has fallen of late years at Selborne.

Burbey took a piece of timber from my orchard, and set a person to turn the water in the pound-field lane. The man, I suppose by the order of Town, did not dig a pit in the hop-garden, but carryed a ditch round by the hedges into Town's mead. A most effectual rain for trial (1·45) fell on the 2nd and 3rd of November. Much white water ran into Town's mead; and so good effect had the contrivance that, comparatively, little came down the street. But behold Parsons came open-mouthed in the morning to complain that your father's expedient had flooded half an acre of his wheat-fallow on the other side of the hedge. However upon examination this outcry, as Parsons himself allowed afterwards, proved to be without reason. I then spoke to Town, who said that Parsons was a poor envious fellow, that could not bear to see him get any benefit from the water; and moreover made a vast merit of admitting the water at all. I see since that some body has stopped the mouth of the drain with a spade!

Before you wrote I had seen accidentally Dryden's 'Hind and Panther.' The poet does not compare *martins* to *Dutchmen*; but *swifts* to *Swiss*, on account of their bulk. What a strange, long-winded allegory has the Laureat†

* Author of *Observationes de Aere, etc. Vide* Letter LX. to Barrington.
† William Whitehead.

made of the three species of Hirundines? little superior, some beautiful passages excepted, to the fables of despised Ogilby. Many thanks for your quotation from Taylor the water-poet, which is very quaint, and comical: and in particular for the Latin one from Gassendus, because it so exactly describes my case. My head to this day is full of the lessons of M. and E. Barker.

We are likely to lose poor W. Dewey: he is afflicted with a terrible asthma, and is in the last stage of a dropsy. He will be missed in his various capacities, and as an honest, blameless man.

The knitter has finished two pairs of stockings, a fine ribbed one, and a pair as thick as a jack-boot.

On November 13th and 14th my Barometer was at 30·2 and 30·3.

Pray has your father received my long ann. dividend up to Midsummer 1782? When opportunity offers pray send me down one pound of Mr. Todd's 14s. green tea. As Mrs. White and I were returning from Fyfield on Novr. 1st, we saw several house-martins playing about under the chalk cliff at Whorwel: the air was frosty, the sun warm.

Mrs. J. White joins in respects.

Mrs. Clement and child go well.

Your loving friend,
GIL. WHITE.

To Miss White.

Selborne, Dec. 14, 1782.

Dear Molly,—I must trouble you again to desire your father to send us, when he has an opportunity to buy it, half an hundred of good salt fish, to be sent down by Earwaker the carrier who puts up at the Castle and Falcon in Aldergate street. When the fish is sent, pray acquaint me by letter of the price, that I may settle with my neighbours.

Now you mention a falcon, one of the keepers of Wolmer forest lately shot a peregrine falcon, a noble bird, which weighed 2 pounds and a half, and measured from wing to wing 42 inches.* I had one sent me before in 1766, which I sent to Mr. Pennant.

We have now, what you would little expect, 26 highland soldiers quartered in this parish; 14 in the street, and 12 at Oakhanger. They belong to the 77th regiment, and were embarked in the S. of Ireland in order to have attended Ld Howe to Gibraltar: but a cross wind drove them to the back side of Cornwall, and so to Ilfracombe on the N. of Devon, where they were landed. These sans-breeches men make an odd appearance in the S. of England.

Mrs. J. White desires that when you write to Miss Isaac, you would mention her son's having set up as a surgeon at Salisbury; because she understands that our cousin boards with a Physician.

We have had much dry weather of late, aud little more than half an inch of rain since the first week in November. I am to be sponsor to my great niece Clement.

Yours affectionately,

GIL. WHITE.

* This incident is duly noted in the *Naturalist's Journal*, where it is added that "the plumage answers well to Pennant's 'Brit. Zoology,' 4to, vol. i. p. 156." Peregrine falcons are still sometimes seen at Selborne; Mr. Parkin, the present owner of "The Wakes" there, recently saw one flying overhead and striking at a heron. A noble sight !

CHAPTER IV.

To the Rev. R. Churton.

Seleburne, Jan. 4, 1783.

Dear Sir,—Your long and communicative letter of Dec. 16th gave me much satisfaction. After you went away my family became very large for the rest of the summer. I had with me my brother Thomas White, and daughter and two sons, my sister Barker from Rutland and her two youngest daughters, and at times my nephew J. White son of Mrs. J. White, who is just settled at Salisbury as a surgeon, being invited by some friends who seemed perswaded that there was an opening. My nieces, Barkers, especially the eldest of the two who is 22 years of age, have (I speak as a foolish uncle) very fine fingers, and play elegantly on the harpsichord. These maidens entertained us day after day with very lovely lessons from Niccolai, Giordani, and several other modern masters, in a very agreeable manner. But I find, as I grow old, that music, though very sweet and engaging at the time, yet occasions very unpleasing sensations afterwards. When I hear fine lessons I am haunted with passages therefrom night and day, and especially at first waking, which by their importunity give me more pain than pleasure : airs and jigs rush upon my imagination, and recur irresistably to my memory at seasons, and even when I am desirous of thinking of other matters. The following curious quotation strikes

me much by so well representing my own case, and by
describing what I have so often felt, but never could so
well express. "Praehabebat porro vocibus humanis, instru-
mentisque harmonicis, musicam illam avium : non quod aliâ
quoque non delectaretur ; sed quod ex musicâ humanâ
relinqueretur in animo continens quaedam attentionemque
et somnum conturbans agitatio ; dum ascensus, excensus,
tenores, ac mutationes illae sonorum, et consonantiarum
euntque redeuntque per phantasiam : cum nihil tale re-
linqui possit ex modulationibus avium, quae, quod non
sunt perinde a nobis imitabiles, non possunt perinde inter-
nam facultatem commovere."—*De vitâ Peireskii per Gassen-
dum.*

I am glad that you met with the Star-sluch in Cheshire,
after you had examined the *Tremella nostoc* in Hants. Not
that I had any doubt myself but that the former was a
vegetable, but because I met with intelligent people who are
still perswaded that this substance is a mass of indigested
food cast-up out of the stomachs of crows ! and some have
told me that they have distinguished the limbs of frogs
among it ! As to a star-sluch growing on the bough of
an oak, this must have been a matter of accident. The
seeds of all *fungi*, you know, are lighter than air, and
therefore float about in it ; and vegetate only when they
happen to fall on a proper *nidus.*

Dr Chandler seemed a good deal chagrined about the
behaviour of his prime minister. If he had not come home
just in time, a *bern* would have been born unto him in
the vicarage. Sim Etty, tutored by the D^r, runs about
the village, and repeats to every one he meets, with great
vehemence :—"Mulieri ne credas, ne mortuae quidem."
Charles Etty is at the Nore aboard the Duke of Kingston,
and is expected every day at Spithead ; from whence he
is to make a visit here for a day or two before he sails
for India.

I thank you much for procuring Mr. Hampton's pamphlet, which you will please to leave at my brother's. You will, I hope, make yourself known to him; I have mentioned you to him. You will see a roomy shop, well furnished, with old gent[lemen] in leathern doublets. Timothy the tortoise would make but a poor king: he would be so slow in his motions as to be but a king Log at best; and an alert enemy would deprive him of half his dominions, before he could awake from his profound slumbers.

I will take care of your *Rex Platonicus*, and hope I shall bring it you at Easter. My brother Thomas opened several of the barrows on our down in the summer, but found nothing. Now you talk of last summer, it was a strange summer indeed! Nothing like it, I believe, has befallen since the year 1725, when it rained every day, except about ten in July, from March 29th to September 29th; but then the first part of said year was very dry. In 1782 the rain that fell at Selborne was 50 in. 26 hund.! and of this the greatest part came in the first 9 months; for Octr., Novr., and Decr. were comparatively dry; Decr. afforded only 0 in. 91 h. I would have you dine with my brother Ben in Fleet street: he dines always about three o'clock. If you would call some morning at my brother Tho. White's at South Lambeth, just beyond Vauxhall turnpike, he would be glad to see you. It is a pretty walk from town to S. Lambeth! If you will go there and dine * . . . .
. . . Sunday, you will meet both families; for they both live * . . . .

> Say what impels amidst surrounding snow,
> Or biting frost the crocus-bloom to glow :
> Say, what retards, amidst the summer's blaze,
> Th' autumnal bulb, till pale declining days ?

* Letter imperfect.

The God of seasons, whose pervading power
Controls the sun or sheds the fleecy shower ;
He bids each hast'ning flower his word obey,
Or to each lingering bud enjoins delay.*

I am with all due esteem,
Your most humble servant,
GIL. WHITE.

Neighbours are all well. Mrs. J. White joins in the good
wishes of the season. If you know any gentlemen in or
round Salisbury pray mention my nephew John White
the surgeon to them, who he is and what he is. D^r
Chandler has been very kind that way. All beginners in
any calling stand much in need of such good offices.

To Mary Barker.   Selborne, Jan. 22, 1783.

Dear Mary,—It is full time that I should acknowledge
your late obliging letter; and return you and your Mother
and sister my best thanks for the agreeable visit that you
made me in the autumn. I have only to regret that you
could not consistently with the respect that was due to
other relations extend it out to a much greater length.

As to music, your lessons, and those of your sister gave
me wonderful delight. I retain still a smattering of many
passages on my memory, which I sing to myself when I am
in spirits. Indeed I am often too much affected with
musical harmony, especially of late years. The following
curious Quotation strikes me much by so well representing
my own case; and by describing what I have so often felt,
but never could so well express†. . . .

When I hear fine music, I am haunted with passages

* 'On the Early and Late Blowing of the Vernal and Autumnal Crocus.'
The verses given above were sent to John Mulso, who had been the recipient
of so many of his friend's compositions, under the signature of "Nobody."

† Here follows the passage from Gassendi, sent him by Miss White, and
quoted in the letter to Mr. Churton of January 4th, 1783.

therefrom night and day; and especially at first waking, which by their importunity give me more uneasiness than pleasure, still seizing my imagination, and recurring irresistably to my memory at seasons; and even when I am desirous of thinking of other matters. . . . Yet notwithstanding all these fine things, I would give six pence to hear you two maidens perform the *wopses*, the lesson with the jig, and that with the lovely minuet, &c., &c.

The letter from *Nobody* puzzled the Mulso family for a long time. At first they suspected me: but the strange, unknown hand, the London post-mark, and some other circumstances threw them all out; so that to put them out of doubt, I was forced to own the imposture, and to acknowledge that you were accessory. Mrs. Clement held her Xtening lately: I was Godfather; and we named the child Jane. Mr. Charles Etty came in this morning from Spithead, where his ship, the Duke of Kingston, is lying at anchor in readiness for sailing soon. This young gentleman says that peace is the general talk: so that he supposes they may possibly sail with a white flag, and without any convoy at all. We have had all this winter 26 High-landers of the 77th regiment quartered in this village, and at Oak-hanger: where though they had nothing in the world to do, they have behaved in a very quiet and inoffensive manner; and were never known to steal even a turnip, or a cabbage, though they lived much on vegetables, and were astonished at the dearness of Southern provisions. Late last night came an express ordering these poor fellows down to Portsmouth; where they are to embark for India; near 100 of them aboard Charles Etty's ship. Uncle Harry writes word that he hopes his son Charles will have a commission soon.

With all due respects I remain

Your affectionate uncle,

GIL. WHITE.

Several of our soldiers came from Caithness.

To Miss White.   Seleburne, Feb. 7, 1783.

Dear Molly,—As the spring begins to advance, and as we propose now being with you about the first week in March, we can hardly wish for half an hundred of salt fish so late : and Mrs. Yalden and Mrs. Etty, I find, are of the same mind. We must therefore desire your father to send a note to his fish-monger to stop his hand.

Having expected the Rector of Faringdon for some time at my house, I could not so well say when about we should endeavor to get to town; but as he has been here, we shall hope to be at liberty as above mentioned, and should be glad to know if that season would be convenient.

Mr. and Mrs. Etty have been very uneasy about Andrew, who still continues in a very poor state, Mr. E. lately has dreaded that a palsy would be the consequence. Mr. Webb pronounced last night that the disorder is St. Vitus's dance, as yet in a small degree. Mrs. E. and Mary went to London last Tuesday : Charles lies still at Spit-head. It was a pity that the parcel was sent by the waggon, because it was detained a fortnight at the warehouse till the tea has lost some of its flavour. Mrs. Clement is at Newton : she has removed her girl to Norgates in Newton-lane, being displeased with her Alton nurse. Mrs. Clement is not very well, and has got a sore throat. Mr. Yalden has some what of the Gout.

" Look upon the rain-bow, and praise Him that made it : very beautiful is it in the brightness thereof."

Ecclus. xliii. 11.

> On morning or on ev'ning cloud impress'd,
> Bent in vast curve, the wat'ry meteor shines
> Delightfully, to th' levell'd sun oppos'd ;
> Smit with the gaudy scene, th' unconscious swain
> In vacant mood gazes on the divine
> Phænomenon, gleaming o'er th' illumin'd fields,
> Or runs to catch the treasure which it sheds.

Not so the sage, inspir'd with pious awe ;
He hails the federal arch ;* and looking up,
Adores the God, whose fingers form'd this bow
Magnificent, compassing heaven about,
With a resplendent verge, 'Thou mad'st the cloud,
Maker Omnipotent, and thou the bow ;
And by that covenant graciously hast vow'd
Never to drown the world again : henceforth,
Till time shall be no more, in beauteous train,
Season shall follow season, day to night,
Succeed '—inspir'd, so sang the Hebrew bard.†

   * Genesis ix. 12–17.   † Moses.   MIMO MILTON.

The end of Jany. and this month have been very wet; so that I fear the springs will get very high, and that the season for our spring-crops will be bad, especially if harsh weather succeeds at once.

How will your Cellars come off? When do the young men‡ go to College?   Your affectte. uncle,

GIL. WHITE.

Rain,
   Jany. 27th, 1·05 ; 30th, ·62 ; Feby. 5th, ·80 : 7th, ·75.

Mr. Dennison is chosen to Holiburn [School] in the room of Mr. Willis : poor Mrs. Robinson, who has 10 children, made what interest she could for her husband, who at present is a Navy-chaplain in the W. Indies : she got two votes, her opponent three. There are but five feofees, one of them a broken blacksmith.

Undated, but of this time, is the following to the same niece, who had on Feb. 10th, 1783, thanked her uncle for his verses on the Rainbow and "as we are no poets," sent him an extract from Thomson's 'Seasons' (Spring, ll. 203–217), in which occur—

    "Here, awful Newton, the dissolving clouds
      Form, fronting on the sun the showery prism."

‡ Miss White's brothers, who were at Oriel College, Oxford, where six of Gilbert White's nephews were now in residence.

To Miss White.   Selborne.

Dear Molly,—As you seemed to be so versed in Milton, I send you an other quotation applicable to the occasion mentioned in my former letter.

> . . . " He . . . . .
>
> . . . .
>
> . . . .
>
> . as an evening dragon came,
> Assailant on the *perched roosts*
> And *nests* in order rang'd
> Of *tame villatic fowl.*" . . . .

Here, my dear niece, your Newtons and your indexes will not avail you.   Yours affectionately,

                                   GIL. WHITE.

To Miss White.

                              Selborne, Feb. 20th, 1783.

Dear Mrs. Mary,—I return you many thanks for your very entertaining letter; and for your method of making of mice; which is a receipt that no family ought to be without; and especially my family, as we have not had a mouse in the house for months.

We have had a strange wedding lately. A young mad-headed farmer out of Berks came to marry farmer Bridger's daughter, and brought with him four drunken companions. He gave two guineas and a crown to the ringers; and came and drank with them, and set all the village for two days in an uproar. Poor dame Butler, hearing that her son was fighting, fell into fits, and continued delirious two days. These heroes, after they had drank all the second day at the Compasses, while a dinner dressing at the great farm at Newton was spoiled, went up at last, and ranted and raved so, that they drove the two Mrs. Hammonds (one of whom is the bride's eldest sister) up into their chambers through fear. At six in the evening they took the bride (who wept a good deal) and carried her away for Berks. The common

Waller&Cockerallph.sc.

Mary White

people all agree that the bridegroom was the most of a
gentleman of any man they ever saw. He told the folks at
the inn that whenever the next sister was married, he
would come and spend ten guineas.

Our crocus's begin to look gaudy.

By the accident that happened to Miss Woods's suit of
cloaths, which was entirely consumed, Geo. Fort's chamber
and furniture sustained much damage, and his house was in
danger.

In riding from Alton to Selborne Mrs. Etty had a fall;
but being light, she was little hurt. The mare fell, having a
stone in her foot. Poor Mrs. Hoar of Nore-hill dyed this
morning!

Pray present my respects to your father, and tell him I
was much concerned, on casting up our account, when I
came to find that by the sudden rise of stock, he had over-
purchased himself in my long ann., and had laid out more
than £20 of his own money. If he pleases to have his
money before I get to town, I will send him a draught
on brother Ben. We talked of setting out on March the
4th, but now find it more convenient to defer our journey
'til Mar. 11th.   Y^r loving Uncle,

GIL. WHITE.

(Barometer)
Therm. Feb.   9, 83   28–2¾
Therm.   „   16 „   30–2½
Rain in Jan., 4·43.
Rain from Jan. 24 to Feb. 14, 5·22.

The following letter was written to Edmund, son
of Benjamin White, at Oriel, who took Orders and
became Vicar of Newton Valence, and consequently
his uncle's neighbour, after the death of his uncle,
the Rev. Richard Yalden :—

To the Rev. Edmund White.

Selborne, Feb. 28, 1783.

Dear Nephew,—I could wish that you would make it
a rule to read aloud to yourself every day some portion
of Scripture or the common prayer, though ever so short,
and that you would sometimes also read before a judicious
friend. But at the same time read plainly and unaffectedly,
and do not aim at anything theatrical or fine: but only
attend well to what you read, and your own good sense and
ear will tell you at the time how to modulate your voice,
and lay your accents justly, according as you are affected by
what is before you.

Mr. and Mrs. Yalden, who are just gone, and Mr. and
Mrs. Clement now in the room, join with Mrs. J. White in
respects.

Remember me to White 3^{tius}, 4^{tus}, and 5^{tus}.

I am, dear White 2^{ndus},

Your loving friend,

White 1^{mus}

Give my respects to Mr. Beake, and Mr. Twopenny; and,
when opportunity offers, to the Provost. Mrs. J. White
and I propose to go to town about March 10th.

To Miss White.

Selborne, Apr. 30, 1783.

Dear Mrs. Mary,—If your Father will turn to Baron
Riedesel's travels thro' Sicily, &c. translated by Forster, he
will find, p. 184, some curious particulars respecting the
family of Coypel the painter, then residing in some rank at
Gallipoli, in Apulia. I returned from Oxford on Saturday
last, and was with the Provost,* who has been at Bath.
I am sorry to say that his state of health is still very

* Of Oriel, Dr. Eveleigh.

lamentable. We rejoice to hear that uncle Harry expects more pupils.

You may say what you please, but the original thought of the Epigram is faulty, in making the Naiades subservient to Ceres, who had certainly no influence over water-deities.

For Ulysses's mill-maids, see Pope's Odyssey, vol. 4, p. 75, book 20. We have swallows and house-martins, but no swifts yet. The martins clean out their old nests, and perhaps new-line them. Smith's terrier, I think, should be questioned about the truth of what his master says he saw on the 11th. We want rain: yet the weather is so glorious, that I can hardly wish it to alter. Our rain was Ap. 11th ·51, 23rd ·37: and that is all since we left you. Thermometer 18th 63°, 19th 65°. We have three nightingales singing in my fields! Cucumbers now come by heaps. Mrs. J. White smiles to see you quote Latin so boldly. Miss Ch[arles] Etty and Andrew [Etty] remain much the same. Your account of your neighbour Cr. head of hair made us laugh. Baptist Isaac, it seems, does not leave Oriel; he has changed his mind. I shall be glad to hear from you again soon; and am, dear niece,

Y^r affectionate Uncle,

GIL. WHITE.

Did you see the wonderful Auroras on Sunday night?

To Miss White.

Selborne, May 13, 1783.

Dear Molly,—Mrs. Tom Butler, who is an intelligent person, says that in the time of Queen Eliz. Temple was in possession of the Seymours; and that she does not think that they have any title-deeds older than that reign. The Seymours of that period must probably be the sons of Edward duke of Somerset, Uncle to King Edwd the sixth, and Ld Protector. See new peerage, Vol 1st p. 13, 14, 15.

What owners there were between the Seymours and Pow-letts, I have not yet learned.

In my way to Oxford, having an hour to spare, I stopped at Dorchester, and saw the church. This is a most venerable pile of building, and of a vast length! Some years ago the church was new roofed and repaired by a benefactor, but the large chancel, being in the hands of a half-ruined Impropriator is miserably neglected. In this noble pile I found a Knight Templar in effigy with his legs crossed: so you see he was vowed to the holy land. In Saxon times Dorchester was a Bishop's See: but in the reign of William the Conqueror it was removed to Lincoln: since which this town has been in constant decay, and is dwindled down to a paltry village. I saw a gentleman lately, who says, many of the Roman coins found at Dorchester were made at Birmingham. For an account of Dorchester in Oxford-shire, see Camden's 'Britannia.'

We have felt many smart frosts lately, and on May 7th had snow and hail-storms: but now the weather is very summerlike.

Not long since I bottled out some very fine raisin-wine. The next morning before I was up Thomas came and told me that he thought that the wine had fermented and broke some of the bottles; for a stream of wine ran from under the vault-door. To this I replyed like a philosopher, that it was impossible that the fermentation could be come to such a degree in so short a time. Thomas came up again, and told me that the stream smelt also of brandy or rum. This account confirmed my first suspicions. So I got up and went into the cellar: when, woe is me, the shelf was fallen down and—*caetera desunt*. We had no *swift* 'til May 4th, and then only a pair: but now this morning, May 13th, we have five or six pairs.

Y^r loving Uncle,

GIL. WHITE.

My tulips are blown out, but are as yet small : but you must observe that the cups of those flowers keep enlarging almost as long as they continue in perfection. Andrew Etty is worse, and Mrs. Etty not well.

On July 12th, 1783, Mulso, referring to the little sets of verses which his friend had recently composed, wrote :—

"That I thoroughly admired all your lines I think I told you; but I communicated them to many friends, and you gained just as many admirers; and even transcribed them all to my sister Chapone; now I shall observe that she liked the 'Rainbow' least, and Mr. Nott liked it best; I do not know which flattered your opinion most. But all the Pieces were much admired. I did not know which to prefer; they seemed to be professedly imitations of several stiles of poets, and in that they seemed equally just. You must not wonder that you did not hear from my daughters upon it, to whom they were directed. They could hardly be answered but in their own way, and that they did not dare to attempt."

To Miss White. Selborne, Aug. 8, 1783.

Dear Molly,—Whenever your brothers want a walk, pray desire them to go to the nursery-garden, late Shields, and to buy me two dozen of the roots of the Dog's toothed violets, and to pay for them, and then to bring them to you, who will, I trust, bring them to Selborne when you come.

I once wished to have had your father here early in July; but since I no wise regret that I was disappointed. For the weather turned out, the month through, violently hot, so that nobody could stir about. Now your father, I know, when at Selborne, loves to bustle about, and to have somewhat in pursuit, and to take walks before dinner. But last month I could neither walk or ride; for the flies tormented

the horses so that there was no peace. Mr. Barker set out
from Lyndon on horseback last Monday, and arrived here
on Wednesday evening without the least complaint or
fatigue. The distance is 118 miles, which he rode with
ease, besides walking 4 or 5 miles at every baiting-place
in his boots, while his horses were eating their corn. He
has still a streight belly, and is as agile as ever; and starts
up as soon as he has dined, and marches all round Hartley-
park. This morning Mr. B. ran round Baker's hill in one
minute and a quarter; and Sam [Barker] in somewhat less
than a minute.

Sally Dewey, though married, is still willing to be employed in my service: I have her now, and have retained
her for the autumn: so there will be no need of troubling
you to bring down a maid.

When my house is at liberty I shall be glad to see you
all. I was in great hopes of seeing Mr. Brown * and Co.
again, but they returned by Sarum and Fyfield without
making much stay at Southampton. Mrs. Brown is better.
Mrs. Etty and Co. are at Priestlands: Mr. and Mrs. Yalden,
for this week at Funtington.

Pray buy me, and bring down when you come

> 1 pound Hyson-tea 14s.
> 1 p^d green tea 10s.
> 1 p^d of coffee.
> ½ p^d of chocolate.
> 2 dozen Dogs toothed violets.

Y^r loving uncle,

GIL. WHITE.

Thus far we have had sweet harvest weather.

Is Colman's translation of Horace's art of poetry † well
received? Mrs. Etty returns to-morrow sennight.

* Mr. Edward Brown, of Stamford, Lincolnshire, husband of Gilbert
White's niece, Sarah Barker.
† This work of George Colman the elder was published in this year (1783).

Gilbert White's nephews, Samuel Barker and John White, visited him early in August. From Selborne they went to spend the day with John Mulso, who wrote on August 26th, 1783 :—

"I thank you for sending me over two such agreeable and accomplished young men. They put me in mind of the times in which we used to take our airings together, and seek for every high hill and every green tree. I have confessed to them that I am broken-winded; they have hinted the same of your horse, but not of yourself; but tell me that you are well and in spirits. My sister Chapone is here. She seemed alarmed that I had told you that she did not like your imitation of Milton; that I did not, nor I could not, justly say; but I said that she liked it the least of the three, and for this you have assigned, perhaps, the just reason. We all here love to talk of you and your place."

The curious long-continued haze described in the following letter was noticed at some length in 'The Natural History of Selborne,' Letter LXV. to Barrington. The cause, which was unknown to the writer, was presumably the fine dust floating in the atmosphere, following a violent volcanic eruption. So far Gilbert White's commentators are agreed, but none of them have traced this occurrence to its origin, which was undoubtedly a tremendous volcanic outbreak in Iceland, in June, 1783, in or near the Skaptarjökull on the north-west border of the Vatna-jökull.

To the Rev. R. Churton.   Seleburne, Aug. 20, 1783.

Dear Sir,—Though my house is full of company, yet I must no longer delay to answer your agreeable and

intelligent letter from Williamscot. Poor Mrs. Etty has
been a great sufferer both in mind and body, having paid
a long attendance on her son Andrew, who languished from
spring to midsummer, and then dyed of a slow decay. What
added to the affliction was, that Miss Charles Etty* was lying
all the while under the same circumstances at Winchester,
and dying first, was brought to this place; so that I had
the sorrowful office of burying these two young people, the
one on one Saturday, and the other on the following.
Charles Etty has not been heard of since he sailed for India
in March; but the papers mention the Duke of Kingston
(his ship) having called at the Cape Verds in April, all well.

We have experienced a long summer, with intense heats,
little rain, and no storms. But what has been very extraordinary, was the *long-continued haze*, extending through
this island, and, I think, through Europe, attended with
vast honey-dews, which destroyed all our hops, and lasted
more than a month. Through this *rusty coloured* air, the
sun, "shorn of his beams," appeared like the moon, even
at noonday. The country people looked with a kind of
superstitious awe on the red lowering aspect of the great
luminary, "Cum caput obscurâ nitidum ferrugine texit."
And I have no doubt, but that the unusual look of the sky
at Cæsar's death, mentioned both by historians and poets,
was somewhat of the same kind.† As I love to trace
natural appearances, I desire to know if you saw a very
large luminous meteor traversing the sky from N.W. to S.W.
on Monday even Aug. 18th about 9 o'clock. Pray hunt for
star-sluch, because several intelligent people, one at present

* *i.e.* daughter of Charles, elder brother of the Rev. Andrew Etty.
† Henry White of Fyfield records in his Journal at this time—
"1783, 17th July. The sun sinks away every day into yᵉ blue mist about
5 p.m., and seems to set behind vast clouds.
"19th. Air seems clearer from yᵉ late blue thickness, which has been so
very remarkable that the superstitious vulgar in town and country have
abounded with the most direful presages and prognostications."

in this house, stare and wonder when I advance that the matter is vegetable; and Dr Chandler in particular shakes his head, and asserts that the mass is frogs thrown up indigested. But I beg to know why crows are not sometimes crop-sick, and have not weak digestions in Hants (yet we have no such appearance) as well as in Cheshire. Apply a magnifying glass to the substance, and try to discover the seeds.

I return you thanks for Hampton's pamphlet, and am indebted to you whatever it cost. The notices concerning Wolmer-forest in the 'Gentleman's Magazine,' came, I conclude, from Dr Chandler, whose extracts from the Worldham Register are genuine. We have this year a most lovely harvest, much corn—but no hops. Our fruit is well ripened, and grapes very forward.

You pay an high compliment to my 'Crocuses,' but were not aware that it will bring more lines on your back. Read them, as little exercises, made last autumn for the use of my nephews (for such they really were), and then you will give them all reasonable allowances. Some weeks ago Dr Chandler was at Portsmouth; but we have not seen him. The D^r does not seem disposed to settle. May I presume to send my humble respects to Dr Townson, whom I have sometimes seen, a long time ago, at Magdalen College. Sportsmen expect a vast breed of game this season. Pray be so good as to favour me with a letter at your leisure. Mrs. J. White joins in respects. I am

Your obliged servant,

GIL. WHITE.

I am glad that you are pleased with the passage from the life of Peireskius, and that you, as well as myself, have been haunted with passages in music.

If you will look in the 'Gentleman's Magazine' for June 1783, you will find, under article "Metamorphosis," a copy

of verses written by a poor dear Oxford friend long since
dead, who was pleased, about 35 years ago, to make himself
merry with my attachment to gardening.*

A HARVEST SCENE:
AFTER THE MANNER OF THOMSON.

Wak'd by the gentle gleamings of the morn,
Soon clad, the Reaper, provident of want,
Hies chearful-hearted to the ripen'd field ;
Nor hastes alone ; attendant by his side
His faithful wife, sole partner of his cares,
Bears on her breast the sleeping babe ; behind,
With steps unequal, trips the infant train.
Thrice happy pair, in love and labour joined !
　All day they ply their task ; with mutual chat,
Beguiling each the sultry, tedious hours :
Around them falls in rows the sever'd corn,
Or the shocks rise in regular array.
　But when high noon invites to short repast,
Beneath the shade of shelt'ring thorn they sit,
Divide the simple meal, and drain the cask :
The swinging cradle lulls the whimp'ring babe
Meantime ; while growling round, if at the tread
Of hasty passenger alarm'd, as of their store
Protective, stalks the cur with bristling back,
To guard the scanty scrip and russet garb.

To Miss White.
　　　　　　　　　Seleburne; Aug. 27, 1783.

Pray Mrs. Mary, did you observe the curious and resplendent meteor on Monday Aug. 18th a little after nine in
the evening, which alarmed the Country people much, who
all agree that it was a Fire-drake. Our shutters were shut
to the N.E. so that we saw nothing of the matter; but
Mrs. Clement and Co. who were at supper with their
windows open, ran out and had a fine view of this
Phænomenon. Sam Barker was at that instant talking
with the Ostler at the Inn in Leatherhead, where he saw
it as well as could be expected in the crouded horizon of a
town.

* *Vide supra*, vol. i. p. 50.

Mr. Barker, the father, is the same agile being that he used to be: he rises at six, and cannot sit still; but starts up the moment he has dined, and runs away to Hawkley Hanger, or King John's Hill. He rode in three days from Lyndon through Oxford to this place 118 miles; and at every dining place, while the horses were baiting, walked four or five miles in the heat in his boots with his wig in his hand.

As soon as I read your translations I knew at once that the first must belong to the Scotch Bishop Gawen Douglas: as to the second, had I supposed that Chapman had ever translated Virgil, as he did Homer, I should from the numbers have imputed it to him; but not recollecting such translation, I suspect it may be done by Phaer and Twyne.

Pray desire your father to receive my midsummer dividend, and to bring me down thirty Pounds, when he comes, and also to remember to charge me for the two hams which he procured for me in the Spring. Be sure remember the Dog's toothed Violets. We have suffered much in our grass-grounds and gardens from drought, but on the 24th and 25th instant we had fine showers. Wasps began to abound and to be very troublesome; but I have employed the Boys to destroy seven nests, and have set my nephews with lime twigs to catch many hundreds more, so that at present the felon race is much lowered.

We have also set bottles with treacle &c.

> They by th' alluring odor drawn, in haste
> Fly to the dulcet cates, & crouding sip
> The palatable bane : joyful thou'lt see
> The clammy surface all o'er strown with tribes
> Of greedy insects, that with fruitless toil
> Flap filmy pennons oft to extricate
> Their feet, in liquid shackles bound, till Death
> Bereave them of their worthless souls. Such doom
> Waits luxury, and lawless love of gain !

Mrs. Etty and Co. returned last Friday all well.

Pray bring me a 1s. box of Greenough's lozenges of Balsam of Tolu.

We all join in respects.

> Y^r loving and affecte Uncle,
>
> GIL. WHITE.

From the *Naturalist's Journal*—

"Sep. 13. Began to mend the dirty parts of the Bostal with chalk.

"Brought down by brother Thomas White from South Lambeth and planted in my borders

"Dog toothed violets. Persian Iris. Quercus cerris. Double ulmaria. Double filipendula. Double blue Campanula. Large pansies. Double daisies. Hemorocallis. White foxglove. Double wall-flower. Double scarlet lynchis.

"Planted from Mr. Etty's garden a root of the *Arum dracunculus*, or *Dragons*, a species rarely to be seen; but has been in the vicarage garden ever since the time of my Grandfather, who dyed in spring 172$\frac{7}{8}$.

"Oct. 4. This day has been at Selborne the honey-market: for a person from Chert came over with a cart, to whom all the villagers round brought their hives, and sold their contents. Combs were sold last year at about 3$\frac{3}{4}$d. per pound; this year from 3$\frac{3}{4}$d.–4d."

To Miss White.

Fyfield, Nov. 6, 1783.

Dear Niece,—When I desired you to procure me some fit stockings, I was afraid that I had enjoined you a troublesome task; however I return you many thanks for the pains you have taken, and would wish to have some knit, provided you think there is any probability of success; but I once bespoke some in that way, and when they came they were long enough for the Giant at Weyhill.* We are much pleased to find that your moving household went on so smoothly, and that you had such fine weather. You must not expect to feel yourselves quite at home at your new house at first: some matters will not be perhaps so convenient as in your old place. You know I love to see fine houses, and furniture: therefore I took a walk yesterday to Mr. Gauler's,† where I was much entertained. The Lady shewed us the whole house. The offices are all under ground, and the kitchen is two feet lower than the cellars! Last night we had grand fire-works. Roman candles, serpents, and sky rockets.

Suppose you get me two pairs of woven stockings of the size and colour desired. I should wish to have them thick and shapely.   Your loving uncle,

Gil. White.

If you can find at last any knit ones fit also, please to get me two pairs.

October 19th, 1783. The *Naturalist's Journal* contains a note of the discovery of two large (carved) stones by the tenant of the Priory Farm "in the space which I have always supposed to contain the South transept of the priory-church," and of the discovery two years before of a vase.‡

* Weyhill Fair was one of the most important fairs in the South of England. *Piers Plowman* mentions it, "At Wy and at Winchester I went to the fair." It was largely attended by buyers and sellers of hops, cheese, etc. On October 11th, 1783, Henry White of Fyfield records in his Journal, "Hops, none from Selborne and very few from that district, few from Farnham, and a very thin shew on yᵉ Hill tho' some Kentish and some old Hops were brt. Best price £11 per cwt. Bought none. Weyhill being yᵉ worst market when they are dear tho' the best when they are cheap."

† Ramridge manor-house, in the parish of Weyhill, Hants, an interesting and beautiful mansion, was built by Mr. Gauler, in or about 1779, from designs by Adams.

‡ These discoveries were recorded in 'The Antiquities of Selborne,' Letter XXVI.

"[Oct.] 26. If a masterly landscape painter was to take our hanging woods in their autumnal colours, persons unacquainted with the country, would object to the strength and deepness of the tints, and would pronounce, at an exhibition, that they were heightened and shaded beyond nature.

"Wonderful and lovely to the Imagination are the colourings of our woodland scapes at this season of the year.

> 'The pale descending year, yet pleasing still,
> A gentler mood inspires ; for now the leaf
> Incessant rustles from the mournful grove,
> Oft startling such as, studious, walk below,
> And slowly circles thro' the waving air.
> But should a quicker breeze amid the boughs
> Sob, o'er the sky the leafy deluge streams ;
> Till choak'd, and matted with the dreary shower,
> The forest-walks, at every rising gale,
> Roll wide the withered waste, and whistle bleak.'
>
> "Thomson's 'Autumn.'"

On November 14th Mulso received a visit from his old friend, who notes in his *Journal*—

"Nov. 14. Winchester. Mr. Mulso's grapes at his prebendal house are in paper bags but the daws descend from the Cathedral, break open the bags, and eat the fruit.

"Looked sharply for martins along the chalk-cliff at Whorwel, but none appeared."

The next day he went home again. He later noted—

"Dec. 5. Fetched some mulleins, foxgloves, and dwarf-laurels from the high-wood and hanger; and planted them in the garden.

"[Dec.] 27. Mr. Churton came from Oxford."

CHAPTER V.

THE following letter mentions a curious recipe for curing the bite of a mad dog :—

To Miss White.

Selborne, Jan. 7, 1784.

Dear Mrs. Mary,—We were much obliged to you for your Liliputian rhimes, which entertained me and Mr. Churton much. If any parcel should be coming down, pray send me half a pound of *break-fast green-tea* at 10s. and half pound of *best tea* at 14s. By the time that I received your favour, you, I trust, received a letter from me desiring your father to send me about 36 pds. weight of *good salt fish*, with the price. I thank God I am better; but have not yet been at Faringdon; and shall, I fear, find it difficult to recover some hardiness, especially as the severe weather is returned, after a rapid thaw. We rejoice to hear that Bro. B. and H. H. W. are so much better. A mad dog from Newton great farm alarmed us much Sunday was fortnight by biting half the dogs in the street, and many about the neighbour-hood. 17 persons from Newton farm went in a waggon to be dipped in the sea, and also an horse. Robert Berriman has lost by illness two horses very lately : and now his cow; which by some strange neglect got into the barn's floor in the night, and gorged herself so at an heap of thrashed

111

wheat, that she dyed what they call *sprung*, being blown up
to a vast size.   These accumulated losses amount, it is
supposed, to full £27!   Your loving uncle,

GIL. WHITE.

Respects as due, and compliments of the season.

On January 29th, 1784, Mulso, who had received
the verses from his friend which have been published
under the title of 'On the Dark, Still, Dry, Warm
Weather occasionally happening in the Winter
Months,' wrote :—

"It was cruel to have a child of yours in my keeping,
and never tell you how I liked it, and how much it had
the features of its father: yet how could I enjoy the
description of calm and occasionally warm weather, when
my very ideas are petrified with cold? Have you remembered anything so severe and so lasting since the year
1740? But as to your poem, I think it super-excellent.
You are quite in your element. I have communicated your
lines to my sister Chapone; she will think with me that
these will be amongst the pieces that you will give one day
to the public."

Here it may be remarked that their author never
published any of his verses, with the one exception
of the poem entitled 'The Naturalist's Summer
Evening Walk,' which is appended to Letter XXIV.
to Pennant, and may truly be termed "super-excellent," and not unworthy to be compared with even
the 'Elegy' itself.   As will presently be seen, however, on one occasion some verses, probably those now
sent to Mulso, were forwarded to the 'Gentleman's
Magazine,' by whom does not appear.

After indulging, like a candid friend, in a little criticism, Mulso continued—

"I do not love to hear of your small inward fever; it was well enough to have a *Hectic heat* when you was young; but I cannot see by your present poetic fury, but that you may be entitled to an honest burning fever, that perspires off in warm verse, and ends in fame to the Doctors and apothecaries; I mean the printers and booksellers, that have watched the crisis and carried your distemper to its end."

To Miss White.   Selborne, Feb. 13, 1784.

Dear Molly,—Having suffered much from the severity of the season, I long for the weather described on the other side of the paper.

According to the Scotch Bishop I am as great a monster as the faithless Æneas; for I

"... sowkit never womanis breist";

having been bred up by hand. Caucasus belongs to the passage you sent me: but one would wonder how the translator could think of lugging in Araby! The "milk unmild" is an extraordinary sort of milk in the second translation; and puts me in mind of *Bristol milk*, a sort of beverage which has destroyed many a morning *whetter*, some of whom I have known. Mrs. J. White and I thank your father and you for your kind invitation. We begin to turn our thoughts towards S. Lambeth; and hope the rigor of the season will soon abate, that we may set out, if convenient to you, on Tuesday the 24th instant. The frost has lasted just 28 days this evening. Last night we had a great snow again, which is much drifted. All friends are well, or mending fast, we hope. I have just received a letter from the Provost of Oriel, who, finding that the extreme west end of Cornwall agreed well with him, never went to Lisbon at all; but has spent several months at and near Penzance, where the kingdom is but eight miles broad. Amidst this world of waters the sea air has been of great service to him, and has restored his flesh: but he is not quite free from complaints.

There have been vast snows in Cornwall, and at Totnes, Devon, whence he wrote, the cold has been intense; he says at 6°.

Y^r loving friend,

G. WHITE.

Lay the verses by 'til I come.*

I have just received your letter by Mrs. Yalden; but not the tea and stockings by Mr. Clement.

The following letter seems to show that Gilbert White, when in town, used to attend the meetings of the Royal and Antiquaries' Societies. There is indeed a family tradition of his shyness on being introduced at these receptions. In a letter to her brother, written on March 22nd (probably 1784), Miss White tells her brother, "My uncle went to the Antiquaries' and Royal Societies last Thursday with Dr. Lort. Mr. Barker's account of the weather was read."

To the Rev. R. Churton.

South Lambeth, Mar. 30, 1784.

Dear Sir,—I take it very kind that you should remember me, when probably I owed you a letter all the while. As I propose to return to Selborne on Friday next, and to set out for Oxford on Easter Tuesday, it does not seem very probable that we shall meet. If you are in London on

---

* Those on the ' Dark, Still Weather,' etc.

a Thursday, I would advise you by all means to attend on the Royal Society and Antiquary meetings in their new splendid rooms at Somerset-house. D^r. Chandler can probably put you in a method of being introduced; if you do not see him, attend in the outer room, between the two rooms, at a *quarter* before *seven* in the evening, and enquire for Dr. Lort, who, I trust, on your using my name, will introduce you to both the meetings, where perhaps you may hear somewhat worth your trouble. The Antiq. Society, I find, is growing very fashionable; for I observed that many Right Honourables were balloted for on Thursday se'nnight. The weather has been dismal and winterlike ever since I left home ; however, I have great advantages in these parts, having a bed at command both in town and country and a carriage to take me to town. Thomas Davis, the bookseller, has just published his memoirs on plays and players, a pleasant book. He has a good stile, and language that no man need be ashamed of, and abounds in curious and pleasant anecdotes. Mr. Etty has heard twice from his son at the Cape of Good Hope; his ship was burnt in the Indian seas, from which he had a miraculous escape, and was carried naked aboard another ship in company; he lost everything. Molly White's rhimes were Norwegian. If you see any lines in the 'Gentleman's Magazine' on such soft weather as I have languished for in vain the spring through, treat them with what lenity you may.

Mrs. J. White joins in respects. If you hear nothing curious at the R. S. or Antiq. meetings, at least you will see two grand rooms and many respectable people, besides Somerset House, a national building as big as three or four colleges!

            I am, with due respect,

               Your most obedient servant,

                     GIL. WHITE.

The next letter, describing his journey home in snowy weather, will be of interest to those who know the locality, and the (now long disused) "hollow lane" from Alton to Selborne; which, to modern eyes, looks a fearsome road to travel, even in summer-time.

To Miss White.

Selborne, April 6, 1784.

Dear Niece,—The many kind offices that we experienced, and the good treatment that we met with from your father and self and other friends at S. Lambeth, deserve an early acknowledgement. The roads were good and without any snow as far as Guildford: but when we came in sight of that town we were surprized to see all the upland grounds covered; and were struck with a chill by the wind blowing over a snowy district. After dinner we mounted Guild-down, and shuddered to behold that Apennine six feet deep in snow under the hedges; and all the sandy country, vales and all, covered over. From Farnham to Alton the scene looked pretty well: but again from Alton to Harteley the snow under hedges was many feet deep; so that had we attempted to have gone earlier in the week, we could not have got along. At the end of the avenue, which was very bad, poor old Selborne afforded a very Siberian view! I hardly knew again my native spot. The down, the hanger, the fields looked very wild and strange!

> "Amidst this savage prospect, bleak and bare,
> Hung the chill Hermitage in middle air;
> Its haunts forsaken, and its feasts forgot,
> A snow-cap'd, lonely, desolated cot."*

* These lines are slightly altered from 'Selborne Hanger,' addressed "To the Miss Batties."

Sunday and yesterday were fine days, during which most of the snow was melted, except under the hedges of the uplands. As I recollect that the havock among the ever-greens about town was great, I was surprized to see how well mine have escaped. My crocus's, retarded a month by the severity of the season, are now in high beauty, and make a glowing appearance. My hepaticas are also fine. There is a good appearance for the bloom of wall-fruit. My Persian Iris's given me by your father, blow well: but stocks and wall-flowers are killed. Thomas has saved many lettuces.

I now see why the late Sir S. Stuart * thought he had pretensions to the Viscountcy of Purbeck. His maternal Grandfather, Sir Richard Dereham of Dereham-abbey, married the Hon. Frances Villiers, the eldest daughter of Robert, Viscount Purbeck, who left no *son*. By this lady Sir Richard Dereham had two children, Sir Thomas Dereham and the late Lady Stuart. Sir Thomas dying unmarried at Rome in 1739, Lady Stuart became sole heir to her Brother Sir Thomas Dereham. So that if Sir S. Stuart had any claim, it was through his mother Eliz. Dereham, who was Grand-daughter to Robert, Viscount Purbeck.

Mrs. Yalden has got a bad cough and cold, caught by her attention to poor Mrs. Etty, who has been in great distress indeed. As to Mr. Etty, I can give but a bad account of him. He got home from Alton on Monday se'nnight, as well as could be expected; but was seized the next day with a delirious fever that deprived him of his reason for some time: however, thank God, he is become very calm, has had some fine sleep, and is much better. A great quantity of blains are now come out over his body and limbs. Mrs. Etty has the satisfaction to have with her Mrs. Stebbing

* Of Hartley House, near Selborne.

and Mr. Charles Etty, her brother; the former of whom
came last Friday, the latter last Sunday. Mr. Litchfield
from Whitchurch is also here. Mrs. J. White joins in all
due respects.

Your loving uncle,

GIL. WHITE.

Sheep and lambs have suffered greatly. Hay is £4 per
ton, mutton 5d. per pound: they talk of 6d.

The poor have suffered greatly for want of work, and the
poor-tax is almost doubled.

I have got a bad cold, but hope to get to Oxford next
week. Little Clement's feet are as bad as ever, both
wrapped in rags; and his eldest sister is come home just
in the same condition. Richard Knight, malster at Faringdon, I hear, has got his ague again. Mrs. Burbey is by no
means well, and cannot use one arm; the cause supposed
to be rheumatic.

A few days after the date of this letter, on April
12th, Mr. Etty was buried at Selborne by Gilbert
White, who registers himself as "curate *pro tempore*."
The loss of this old and intimate friend and neighbour must have been great indeed, especially in such
a retired neighbourhood as that of Selborne. Moreover, he was soon to lose the head of another of "the
three families," Mr. Richard Yalden.

To Miss White.   Selborne, April 19, 1784.

Dear Molly,—Pray present my respects to your father,
and tell him I return him many thanks for the £100 which
he has just paid into nephew Ben's hands: £80 of it was,
I suppose, on his own account, and £20 on Mrs. J. White's:
of this let me hear, to prevent mistakes.

Mr. Charles Etty* is gone to town, I suppose to prove his brother's will. Mrs. Stebbing† had told us that she should stay to the end of this week: but I was surprized to see her in my parlor on Saturday morning, and to hear her say she was come to take leave; and that the chaise was at the door. After a small pause, she told us that she would not conceal from us the reason of this sudden motion: for she was hasting home to consult D͏ʳ Stebbing about the living of Whitchurch,‡ which she had good reason to suppose the L͏ᵈ Chancellor§ was disposed to give him. The Lord Chancellor, it seems, was very nearly related to D͏ʳ Battie, and when a young Counsel used to be almost every day at the D͏ʳˢ house, and consequently intimate with Mrs. Stebbing and Etty: though no acquaintance has been kept up for more than 20 years past, for certain reasons. Mrs. Stebbing doubts much whether her husband will be disposed to take Whitchurch at all, in the first place because the D͏ʳ hates all trouble, and business, even to the writing a common letter: in the next place, he must resign Streatly, which neats him £40 per annum; and the getting into Whitchurch, a second living, and a Chancellor's living, will cost him £70, or £80: so that if he should dye within the year, his family will be a loser. I wish myself he may take it, on Mrs. Etty's account, for reasons too obvious to be mentioned. It is remarkable that by the Battie interest Mr. Etty did enjoy, and D͏ʳ Stebbing may, as it seems, enjoy the desirable and much sought for parsonage of Whitchurch. Mrs. Etty does not now leave Selborne 'til matters are finally settled. Poor Lady, she bears her great loss with wonderful resignation. The next

* Of Priestlands, in the parish of Milford, near Lymington, Hants, a brother of the Rev. Andrew Etty. In the large view of Selborne (in the quarto editions of 'The Natural History') he is represented standing in the foreground. He died in 1797, and was buried at Selborne, by his own desire.   † Mrs. A. Etty's sister.

‡ Held with Selborne by Mr. Etty.   § Lord Thurlow.

taker, it seems, for Selborne, is a Mr. Taylor, a gentleman that has long resided in the New Forest: he is a Londoner by birth.

What between frequent snows, and rain, we have sown no seeds in the garden; and what is worse, the farmers have sowed no spring-corn. Last Wednesday we had a heavy snow all day, which hung in the trees, and covered the ground very deep!! Pray how is it, Mrs. Mary, that the present most *bitter* spring does not at all retard the coming of the summer birds? for the nightingale was heard at Bramshot on March the 30th, the Tuesday before we left you; and a farmer told Mr. Yalden he saw two swallows at Hawkley on April 7th, and again a nightingale was heard at Maiden-dance on April 16th, and again many swallows (perhaps house martins) were seen at Oak-hanger-ponds on April 16th: and a black-cap was seen and heard on April 17th. Timothy begins to move, and to make the mould crumble over his back. On my asking Mr. Yalden whether he thought the farmer a likely man to know swallows, he cryed, " O, yes—for he was a married man." To which I replyed, "that though a very unworthy batchelor, I presumed I knew swallows as well as most married men in England."

Indeed, Molly, I have suffered a great deal since I saw you; but, I thank God, am much better. The death of Mr. Etty in the midst of my illness did not, you may imagine, much assist my spirits. I was not able to go to Oxford. Tell nieces Barkers that I should be very glad to hear *rux the wopses,* and Sam Barker that I hope to write to him soon. Little damage was done in the great parlor by the accident that might have proved fatal to the house, had not both the servants been just at hand.

Mrs. J. White returns you thanks for your kind letter.

---

* Some musical piece. Cf. letter to Mary Barker, Jan. 22nd, 1783.

Pray write soon and mention the business alluded to at the beginning.

Yr loving uncle,

GIL. WHITE.

Crickett, red cabbages, early York, and sugar loaf are in rough leaf.

To Miss White.   Seleburne, May 22, 1784.

Dear Molly,—If your brother Harry was in orders, I could now put him into present pay and good quarters; because the curacy of Selborne seems likely to be at my disposal. The reason of this is, because Mr. Taylor, the probable next vicar, has been here, but does not seem at all disposed to reside. He was very earnest with me to take this church at once: but I told him I could not leave Faringdon abruptly. This gentleman and Mrs. Etty are in treaty about the vicarage-house, which Mrs. E., I find, would be glad to take. Mr. Taylor, I understand, has made connections in the New Forest near Ring-wood; has got a small living in those parts, and expects a better: he is concerned with the Lisle family some how; and is to marry a young lady of that name, as is reported.

Poor Mrs. Etty is now in some perplexity of mind about her son Charles, who wrote her word that he should certainly sail from the Cape in one of Commodore King's men of war. But Mr. King has been now come in to Spithead for some time; yet no young man, nor letter appears: neither by enquiries aboard can they make out the meaning of this disappointment. We all hope that in the interim between his writing and King's sailing, Charles went aboard some Indiaman, and may have called by the way at St Helena.

We have lost poor Timothy, who, being always in a great bustle in such hot weather, got out, we suppose, at the wicket, last Thursday; and is wandered we know not

whither! Thomas is much discomposed at this elopement;
and has

> ". . . made as great a coil as
> Stout Hercules for loss of Hylas.
> He has forc'd the hangers to repeat
> The accent of his sad regret:
> And Echo from the hollow ground
> His doleful wailings to resound."

But to be serious, I should be very sorry to lose so old
a domestic, that has behaved himself in so blameless a
manner in the family for near fifty years. We have leaped
this year from winter to summer at once, like the countries
round the Baltic, without any gradation of seasons. The
Tulips, as soon as blown, gape for breath, and fade.

Yours, &c.,

GIL. WHITE.

	in.	h.
The rain in April was .	3	92
as yet this month only .	0	23!

Pray write soon.

Mrs. J. White thanks you for your letter; and your
father for £10 bank-note.

May 24. Fine showers this morning, but now hot sun
again. No Timothy to be found.

On June 2nd, 1784, Mulso writes :—

"I received your kind letter, signifying your return to
Selborne. You and your family have a turn for improving
every place that you belong to. . . . I hope you will retain
them that are left in your neighbourhood. New friends may
be an amusement, but 'the old are better.' Of Dr Balguy*
it may be said, here is the man that refused a Bishoprick;
and of you, here is the man who refused livings, and served
curacies. . . . I think by the time you leave Faringdon you

---

* John Balguy (1686–1748), an eminent theologian and moral philosopher.

might get my son John in there, and he might get shelter for his head at Selborne, and travel over on Sundays, as you do."

To Miss White.   Selborne, June 12, 1784.

Dear Niece,—I did not mean to tell you that Mr. Taylor would not take Seleburne, because I believe he will. I only intimated that if he does, and does not marry, he may throw it up at any time between Michaelmas next and Michaelmas twelvemonth. So far, you see, his keeping this vicarage is not a settled thing.

Poor Mr. Yalden, within these two or three days, has been ill, very ill. He is better; but we think his state of health very bad at present.

Lady Young,* and her two daughters are now at Mrs. Etty's house. Mrs. Call† was to have been of the party, but was prevented by indisposition.

Mr. Clement has just served thirty-nine people of the parishes of Binsted, Frinsham, &c. with some process in law at the desire of Lord Stawel, because they have carried away all the top and lop of the great fall of timber cut this spring in the Holt. If these reputed culprits do not make restitution, they are all to be prosecuted at the assizes at Winton. One person, who has got a team, has secured to himself near forty stacks of wood. These folks, especially the females, are very abusive, and set my Lord at defiance: for, they say, they can produce the will of one *Alice Holt*, wherein, after bequeathing the Holt to the crown, she has given the lop to the poor of certain parishes, and they threaten also to produce a brass plate, dug up in some church-yard, which is to confirm their claim.

After Timothy had been lost eight days, he was found in the little field short of the pound-field. He had con-

---

* Formerly Miss Battie.
† Formerly Miss Philadelphia Battie.

ceived a notion of much satisfaction to be found in the range of the meadow, and Baker's hill; and that beautiful females might inhabit those vast spaces, which appeared boundless in his eye. But having wandered 'til he was tired, and having met with nothing but weeds, and coarse grass, and solitude, he was glad to return to the poppies, and lettuces, and the other luxuries of the garden.

The more I enquire into the mischief occasioned by the hail-storm the worse I find it! Had not this tempest been confined to narrow limits the whole neighbourhood would have been desolated! You will be surprised to see the heaps of stones that the torrents have washed down!!!

Your loving friend,
GIL. WHITE.

Blowing showery weather for some days. A prospect of much grass. When do you make hay?

We shall hope to see you all in August.

The *Naturalist's Journal* at this time mentions—

"July 17th. Mr. Charles Etty brought down with him from London in the coach his two finely-chequered tortoises, natives of the island of Madagascar, which appear to be the *Testudo geometrica*, Linn., and the *Testudo tessellata* Raii. One of them was small, and probably a male, weighing about 5 lbs.: the other which was undoubtedly a female, because it layed an egg the day after its arrival, weighed $10\frac{1}{4}$ lbs. The egg was round, and white, and much resembling in size and shape the egg of an owl. The backs of these tortoises are uncommonly convex and gibbous." [A reference to "Ray's *Quadrup.*, p. 260" is given]. "The head, neck, and legs of these were yellow. . . ."

At this time the Mulso family from Meonstoke were purposing a visit to Selborne. The usual

request for a guide to meet them at Tisted turnpike, with a comparison of Selborne to the "Bower
of Woodstock," was made.

Mrs. Chapone, also invited, was unable to come.

The entry occurs on—

"July 27th, 1784. Mr. and Mrs. Mulso, and Miss Mulso,
and Miss Hecky Mulso came."

The latter name is noticeable in view of what
followed a little later.

On August 12th, 1784, John Mulso writes his
thanks for "all the kind attentions you paid to me
and mine." He continues—

"Timotheus has been prurient of poetry, and surely now
'his flying fingers have swept the Lyre,' he has shown a
great vivacity, joined with sentiment and solidity. I hope
he will not content himself with speaking *once*, like Balaam's
ass, but will exercise his gifts, having once spoken so well."

The above must be read in conjunction with the
following reply of "Timothy the Tortoise," which, first
published in Jesse's 'Gleanings on Natural History'
(John Murray, 1834), has always hitherto been
assumed to be addressed to Mrs. Chapone, *née*
Miss Hester (Hecky) Mulso. It was, of course,
addressed to her niece, Mulso's second daughter
—then a girl in her twenty-first year—who had
sent to Gilbert White some verses addressed to the
tortoise, after this visit to Selborne.

Mrs. Chapone had for twenty-four years ceased
to be Miss Mulso, a fact which has been deemed

of small importance by some sentimental writers, who have enlarged on "the pathos" of this incident.

Timothy the Tortoise to Miss Hecky Mulso.

From the border under the fruit wall,

Aug. 31, 1784.

Most respectable lady,—Your letter gave me great satisfaction, being the first that ever I was honored with. It is my wish to answer you in your own way; but I never could make a verse in my life, so you must be contented with plain prose. Having seen but little of this great world, conversed but little and read less, I feel myself much at a loss how to entertain so intelligent a correspondent. Unless you will let me write about myself, my answer will be very short indeed. Know then that I am an American, and was born in the year 1734 in the Province of Virginia in the midst of a Savanna that lay between a large tobacco plantation and a creek of the sea. Here I spent my youthful days among my relations with much satisfaction, and saw around me many venerable kinsmen, who had attained to great ages without any interruption from distempers. Longevity is so general among our species that a funeral is quite a strange occurrence. I can just remember the death of my great-great-grandfather, who departed this life in the 160th year of his age. Happy should I have been in the enjoyment of my native climate and the society of my friends had not a sea-boy, who was wandering about to see what he could pick up, surprised me as I was sunning myself under a bush; and whipping me into his wallet, carryed me aboard his ship. The circumstances of our voyage are not worthy a recital; I only remember that the rippling of the water against the sides of our vessel as we sailed along was a very lulling and composing sound, which served to sooth my slumbers as I lay in the hold. We had a short voyage, and came to anchor on the coast of England in the harbour of Chichester. In that city my kidnapper sold me for half a crown to a country gentleman,* who came up to attend an election. I was immediatety packed in an hand-basket, and carryed, slung by the servant's side, to their place of abode. As they rode very hard for forty miles, and I had never been on horseback before, I found myself somewhat giddy from my airy jaunt. My purchaser, who was a great humorist, after shewing me to some of his neighbours and giving me the name of Timothy, took little further notice of me; so I fell under the care of his lady, a benevolent woman, whose humane attention extended to the meanest of her retainers. With this gentlewoman I remained almost forty years, living in a little walled-in court in the front of her house, and enjoying much quiet and as much satisfaction as I could expect without society. At last this good old lady dyed in a very advanced age, such as a tortoise would call a good old age; and I then became the property of her nephew. This man, my present master, dug me out of my winter retreat, and, packing me in a deal box, jumbled me eighty miles in post-chaises to my present place of abode. I was sore shaken by this expedition, which was the worst journey I ever experienced. In my present situation I enjoy many advantages — such as the range of an extensive garden, affording a variety of sun and shade, and abounding in lettuces, poppies, kidney beans, and many other salubrious and delectable herbs and plants, and especially with a great choice of delicate gooseberries! But still at times I miss my good old mistress, whose grave and regular deportment suited best with my disposition. For you must know that my master is what they call a *naturalist*, and much visited by people of that

---

* Mr. Snooke of Ringmer, near Lewes.

turn, who often put him on whimsical experiments, such as feeling my pulse, putting me in a tub of water to try if I can swim, &c.; and twice in the year I am carried to the grocer's to be weighed, that it may be seen how much I am wasted during the months of my abstinence, and how much I gain by feasting in the summer. Upon these occasions I am placed in the scale on my back, where I sprawl about to the great diversion of the shop-keeper's children. These matters displease me; but there is another that much hurts my pride: I mean that contempt shown for my understanding which these *Lords* of the *Creation* are very apt to discover, thinking that nobody knows anything but themselves. I heard my master say that he expected that I should some day tumble down the ha-ha; whereas I would have him to know that I can discern a precipice from plain ground as well as himself. Sometimes my master repeats with much seeming triumph the following lines, which occasion a loud laugh—

> "Timotheus placed on high
> Amidst the tuneful choir,
> With flying fingers touched the lyre."

For my part I see no wit in the application; nor know whence the verses are quoted; perhaps from some prophet of his own, who, if he penned them for the sake of ridiculing tortoises, bestowed his pains, I think, to poor purposes. These are some of my grievances; but they sit very light on me in comparison with what remains behind. Know then, tender-hearted lady, that my greatest misfortune, and what I have never divulged to anyone before is—the want of society of my own kind. This reflection is always upper-most in my own mind, but comes upon me with irresistible force every spring. It was in the month of May last that I resolved to elope from my place of confinement; for my fancy had represented to me that probably many agreeable tortoises of both sexes might inhabit the heights of Baker's

Hill or the extensive plains of the neighbouring meadow, both of which I could discern from the terrass. One sunny morning, therefore, I watched my opportunity, found the wicket open, eluded the vigilance of Thomas Hoar, and escaped into the S^t-foin, which began to be in bloom, and thence into the beans. I was missing eight days, wandering in this wilderness of sweets, and exploring the meadow at times. But my pains were all to no purpose; I could find no society such as I wished and sought for. I began to grow hungry, and to wish myself at home. I therefore came forth in sight, and surrendered myself up to Thomas, who had been inconsolable in my absence. Thus, Madam, have I given you a faithful account of my satisfactions and sorrows, the latter of which are mostly uppermost. You are a lady, I understand, of much sensibility. Let me, therefore, make my case your own in the following manner; and then you will judge of my feelings. Suppose you were to be kidnapped away *to-morrow*, in the bloom of your life, to a land of Tortoises, and were never to see again for fifty years a human face!!! Think on this, dear lady, and pity

Your sorrowful Reptile,
TIMOTHY.

To Miss White.

Selborne, Aug. 16th, 1784.

Did you not, Molly, feel a sharp twinge in one of your arms on the 22nd of July, by sympathy? because on that day the heavy gales, and showers broke off a great bough from your ἀμαδενδρον, the sycamore, and have injured it much! As I have a great regard for you, and all that belongs to you, I am much concerned for your tree.

You know I never dictate to you about your coming to this place: I only injoin you to come before winter, and to stay a long time. We have provided a young person to help while you are here; so you need not trouble yourself to send down your maid. Pray pay Mr. *Almond* for my *two* pairs of stockings: by not taking the money in Fleet street, he has wronged himself: for he returned the *whole* money to Mrs. J. White while we were in town. Desire your father to send me down a good *large ham*:

To bring me down a £30 *bank-bill*:

Bring me a pound of *coffee*:

Half a pound of soft *sealing-wax*:

Two or three quires of *small* writing paper.

Your father, of course, I conclude, will receive my Mid-summer long ann: dividend. Mrs. Etty and family intend to go to Priestlands soon. Charles Etty's fine Madagascar tortoises dyed as soon as they got to Selborne; but not before the female, a very grand personage, had laid an egg. They seem to have been jumbled to death in the boot of the coach.

Brother Ben gathered a puff-ball, *Lycoperdon bovista*, in a meadow at Alton, and brought it to Newton: it weighed seven pounds and an half, and measured in girth, the longest way, 3 feet 2 inches and an half!!

Mr. and Mrs. Mulso, and two daughters have just spent a fortnight with me. You may suppose the company of such very old friends was very agreeable. Mr. and Mrs. M. are very inactive; especially the latter. That Lady and I made a visit to Mrs. Etty in a carriage: and once we went on foot. Mrs. J. White says you need not trouble yourself farther about her hat: she herself will settle that matter. Suppose you send the *Coffee* down by Neps. Ben and Edmund; who, as I understand, are soon to be in Hants. Pray let me hear as soon as you get to Fyfield. I thank your father for paying my insurance. When you come I shall be glad to trace your Essex-tour on the map with you. Tell your father that the wheat-crop in the pound-field (he remembers circumstances) is bad. We have sweet harvest-weather; but the Selborne wheat (especially what was

smitten by the hail) does not ripen together: some part is
very green, while some is dead ripe.

Your loving uncle,

GIL. WHITE.

To Miss White.

Selborne, August 31, 1784.

Dear Molly,—My sealing-wax, paper, and coffee came safe
by my nephews: the latter by mistake was directed for Mrs.
Etty. That gentlewoman had left Selborne just before your
letter arrived: she is gone to Priestlands. We expected Mr.
Taylor last week to come and take possession of Selborne
living: but he came not. He must be inducted soon. We
have had a sad, cold, wet wheat-harvest: much wheat is
housed in a poor cold state. The poor steal the farmers
corn by night:* the losers offer rewards, but in vain. My
quantity of fruit is very great: but nothing ripens. Much
of the wheat of Selborne will be bad, especially what was
smitten by the hail: a great proportion of it will never be
ripe. Lord Rodney the other day came to look at Hartley-
house, which he says he will take, if the trustees will put it
in repair. Mr. Wilmot is threatened with £200 delapida-
tions! The newspapers will tell you of the princely diver-
sions that were carryed on lately at Up-park † for the
amusement of the Prince of Wales, Jack-ass races, and men
jumping in sacks afforded the principal sport! Charles
Etty was there, and many folks from these parts. A hungry
dog came the other night, and standing on his hind-legs,
pulled off as many of my ripe apricots as he could reach:
many of which he swallowed, and many he left half eaten
on the ground. The thief with me was a real dog, *canis*;

---

* How badly off the poor were at this time may be gathered from Henry
White of Fyfield's Journal—

"1784, April 15th. Parish meeting, yᵉ Poor Rates higher than ever, 16!
Rates, 3 more than ever was known before, but 13 last year." [A Rate was 2*d.*]

† In Sussex, near Petersfield.

but there came lately to Mr. Mulso's at Meon-stoke, some two-legged dogs, who stripped his apricot-trees of all their fruit; and the next night carryed away two of his large goose-berry-trees laden with fruit. This fruit was taken for sale at cricket-matches, and ass-races. As Mr. Wools was playing last week at cricket, his knee-pan was dislocated by the stroke of a ball: and at the same time Mr. Webb was knocked down, and his face and leg much wounded by the stroke of a ball. Mr. Yalden is better, and talks much of shooting; but the fields are full of corn: much wheat abroad, and no spring-corn housed. We hope business will not detain you now for any long time: for when once September commences, we may truely say, "Apace the wasting summer flies." Miss Heckey Mulso has written a long letter in verse to *Timothy*; which, with great labour, and pains, he has answered in prose.

Pray let me have early information of your motions.

Hops innumerable, but small.   Y^{rs} &c.,

                                    GIL. WHITE.

The summer has been so bad, that we have had *no white* kidney-beans, and few Cucumbers: the scarlet kidney-beans have born a little.

The hops smitten by the hail are likely to have a good crop: their tops were broken off, but they soon threw out fresh runners. The damage done to the wheat is more permanent. The Crop of Hops at Farnham is vast! We hardly see two fine days together, and very little sun. Miss Butler was lately married to a Mr. Cox, who has taken the parsonage-house at Trotton, near Midhurst. Here is fine after-grass for your father's horses. Please to send me a good large *ham*. Neighbour Hale and I have been both walking in his hop-garden, and in the contiguous one of Spencer, both of which were smitten by the hail: and we both agree that the seeming calamity of the hailstorm has proved a great advantage to each owner. For these plants, being nipped off, have thrown out much *side-wood*, and have produced much *fairer* and *larger* hops than any in the parish; and a vast crop. Quæ. Should not men, when the binds are very strong, nip their tops? we do so with melons, and cucumbers. Hopping begins this day: Sep. 1st.

Rain Aug. 20th—1·61. Rain in the month of Augst., 3·88.

Edmund, and Mr. Clement launched a balloon last night in Mrs. Y[alden's] stair-case, with some success; but it would not succeed abroad.

From the *Naturalist's Journal*—

"Sept. 11. Mr. Randolph, the Rector of Faringdon, came."

To Miss White.   Selborne, Septemr. 24, 1784.

Dear Molly,—We have made a post with one of your arms:* but as the material is young and tender, it will not last long. My wall-fruit is now fine: but the rains, and the bees injure it. We have gathered good grapes once. The latter harvest, and the hops are finely got in. Poor Moses Terry, a bed-rid paralytic, was found dead in his bed this morning: it is to be feared that he strangled himself in his wife's absence with a leathern thong, part of which, being broken, was found round his neck! Tell your father that I have saved all the after-grass of the great meadow for his horses: it is fetlock deep. We hope you will come not long hence. Mrs. Etty has been gone some time: but as you come so late, you may probably see her, before the end of your visit. October is often fine: I hope the next will prove so. November is also often fine: so we hope you will stay with us a good long time. Pray do.

Mr. Taylor has not been here yet to take possession

---

* *i.e.* of the sycamore planted by her.

of Selborne vicarage.*   We expect Sister Harry and Lucy
in a day or two : they are to stop at Alton the first night.
My niece comes here for change of air.   Nephews Ben.,
Edmund and Clement have had a sad wet journey to
Portsmouth, and the Isle of Wight, where they saw naught,
but a dirty inn; not being able to stir out one step.   They
had also squally weather, and were almost drowned with
rain in their passage out.   Mr. Clement has caught a great
cold : the other young men have escaped better than could be
expected.   I was in fear for nephew Ben., because of his
late Infirmities.   O Molly ! you don't tell us of the balloon,
and the ascension of Mr. Lunardi : did it not affect you, to
see a poor human creature entering upon so strange and
hazardous an exploit ?   I wish the newspapers would learn
to talk with a little more precision about thermometers.   In
the late accounts they represent Mr. L. and his apparatus
covered with ice at 35°, three degrees *above* the freezing
point.   One paper says Mr. L. discharged some of his gas,
and *then* the atmosphere was very mild ; not understanding
that the *descent* into a lower region occasioned the mildness!!

Your loving uncle,

GIL. WHITE.

My ham came safe; but had a great escape : for in its
passage down the waggon was robbed of about £30 in value.

To Mrs. Barker.   Selborne, Octr. 19, 1784.

Dear Sister,—From the fineness of the weather, and the
steadiness of the wind to the N.E. I began to be possessed
with a notion last Friday that we should see Mr. Blanchard
in his balloon the day following : and therefore I called on
many of my neighbours in the street, and told them my
suspicions.   The next day proving also bright, and the wind

---

* He took possession on September 26th, 1784, the *Naturalist's Journal*
records.

continuing as before, I became more sanguine than ever;
and issuing forth in the morning exhorted all those that had
any curiosity to look sharp from about one o' the clock
to three towards London, as they would stand a good chance
of being entertained with a very extraordinary sight. That
day I was not content to call at the houses only; but I went
out to the plow-men and labourers in the fields, and advised
them to keep an eye to the N. and N.E. at times.   I wrote
also to Mr. Pink of Faringdon to desire him to look about
him.   But about one o'clock there came up such a haze that
I could not see the hanger.   However, not long after the
mist cleared away in some degree, and people began to
mount the hill.   I was busy in and out 'til a quarter after
two; and took my last walk along the top of the pound-field,
from whence I could discern a long cloud of London smoke,
hanging to the N. and N.N.E.   This appearance, for obvious
reasons, increased my expectation: yet I came home to
dinner, knowing how many were on the watch: but laid my
hat and surtout ready in a chair, in case of an alarm.   At
twenty minutes before three there was a cry in the street
that the balloon was come.   We ran into the orchard, where
we found twenty or thirty neighbours assembled; and from
the green bank at the S.W. end of my house saw a dark blue
speck at a most prodigious height, dropping as it were from
the sky, and hanging amidst the regions of the upper air,
between the weather-cock of the tower and the top of the
may-pole.   At first, coming towards us, it did not seem to
make any way; but we soon discovered that its velocity was
very considerable.   For in a few minutes it was over the
may-pole; and then over the Fox on my great parlor
chimney; and in ten minutes more behind my great wallnut tree.   The machine looked mostly of a dark blue colour;
but some times reflected the rays of the sun, and appeared
of a bright yellow.   With a telescope I could discern the
boat, and the ropes that supported it.   To my eye this vast

balloon appeared no bigger than a large tea-urn. When we saw it first, it was north of Farnham, over Farnham-heath; and never came, I believe, on this side the Farnham-road; but continued to pass on the other side of Bentley, Froil, Alton; and so for Medsted, Lord Northington's at the Grange, and to the right of Alresford, and Winton; and to Rumsey, where the aerial philosopher came safe to the ground, near the Church, at about five in the evening. I was wonderfully struck at first with the phænomenon; and, like Milton's "belated peasant," felt my heart rebound with fear and joy at the same time. After a while I surveyed the machine with more composure, without that awe and concern for two of my fellow-creatures, lost, in appearance, in the boundless depths of the atmosphere! for we supposed *then* that *two* were embarked in this astonishing voyage. At last, seeing with what steady composure they moved, I began to consider them as secure as a group of Storks or Cranes intent on the business of emigration, and who had—

> ". . . set forth
> Their airy caravan, high over seas
> Flying, and over lands, with mutual wing
> Easing their flight. . . ."

Mr. Taylor, our new vicar, has taken possession of S[elborne] living; and I have reassumed the curacy, after an intermission of 26 years! Mrs. Etty rents the vicarage house; but has been gone eight or nine weeks, and does not return 'til winter. Mr. Yalden has gone to Bath in Company with Mr. Budd. Brother Ben. and family are at Newton, but go next week. Brother Thomas has been expected here all the autumn, but is not yet come. Mrs. H. White brought Lucy to my house lately, for change of air: the poor young woman is languid, and has over-grown her strength: but I perceive no bad symptoms. We have apples and pears innumerable, and *very* fine grapes. Mrs. Clement is in a fair way, I suspect, to encrease her family. I wish you joy of your late

grand-daughter, which makes my 41st nephew and niece!
I have very dutiful nieces, that seem disposed to make me as
great an uncle as they can.   Mrs. J. White joins in respects.
I am with all due affection and regard,

Y^r loving brother,

GIL. WHITE.

Sweet autumnal weather! we have had no rain since
Septemr. 27th not enough to measure.   I miss poor Mr.
Etty every day: he was a blameless man, without guile.
His son Charles is in London making interest for an appoint-
ment to India.   His escape off Ceylon was wonderful!

On the same date a letter describing the balloon
journey in identical terms was sent to his niece Molly
at South Lambeth.   He adds:—

" We most earnestly hope to see you soon, and shall rejoice
more at the sight of your post chaise, than if the balloon had
settled on our sheep-down.   Your father's letter on balloons
is very entertaining and the last quotation finely adapted
and happily applied."

His account must have been published, since in
the *Naturalist's Journal* of this date is pasted a
cutting from a newspaper, headed "Extract of a
[the above] Letter from a gentleman in a village
50 miles S.W. of London, dated Oct. 21."   The end
of the balloon journey is thus recorded by Henry
White of Fyfield in his Journal:—

"1784.   Oct. 18. Mr. Wellman came.   He saw on Saturday last p.m. half-past 4, at Rumsey Mr. Blanchard in his
grand air baloon hovering at a great height over the Church,
and soon after saw him descend into a meadow near the
town. . . . Mr. Blanchard was only 3½ hours passing from
London to Rumsey, 75 miles, and was seen passing over many
places particularly from Selborne hill and village."

Mr. Bell (in his edition, vol. ii. p. 156) recounts
that the letter Gilbert White wrote to Mr. Pink
warning him to look out for the balloon was the
occasion of a very ludicrous circumstance. Mr.
Pink, a very respectable yeoman, was on his way to
Alton market, the day after he received the letter,
when he overtook a neighbour, and asked him if
he had heard or seen Mr. White lately. Being
answered in the negative, he said, "Ah, poor man,
he is very far gone indeed!" pointing to his head.
"I had a letter from him yesterday, and what do
you think he said to me, and desired me to do?—he
told me to look out sharp to the north-east between
one and three o'clock to-day, and perhaps I should see
two men riding in the air in a balloon." To which
his neighbour replied, "Then he must be pretty far
gone indeed." Mr. Pink expressed great sorrow,
as he very much respected him. When they came
to Alton Butts, a small open common just as you
enter Alton, a large concourse of people were
assembled together, looking earnestly upwards. Mr.
Pink asked them what they were about; to which
they replied that if he would look over the church
he would see two men riding in a balloon. After
satisfying themselves of the truth of this, Mr. Pink
jogged his companion, saying, "Neighbour, I think
Mr. White is not so far gone as you and I thought
him!"

Living in a neighbourhood where the roads were
so bad Gilbert White was ever interested in road-

Henry White.

[To face p. 138, Vol. II.

making, like his grandfather, who bequeathed money for this purpose. Among the few anecdotes of him, which the writer has been able to collect in Selborne, is the story that when the Naturalist was seen approaching, the village children used to put stones in the ruts, and receive from him pennies for their diligence. The following letter refers to this subject:

To Miss White
  At the horse and jockey
    Newton-lane end.   Novr. 23rd, 1784.

Dear Molly,—When you come to Newton-cross I wish you would turn short on the left, and so go along Newton-lane, where the quarter,* I think, is made very safe. After you have passed the N. field, turn down the N. field-hill-lane, which has had much labor bestowed on it, and is, I trust, now very safe also.

If the latter should prove as I expect, having never been used for quarter before, you may say—

> ". . . juvat ire jugis, qua nulla priorum
> Castaliam molli divertitur orbita divo."

                                  Y^r loving Uncle,
                                    GIL. WHITE.

On the same date as the above letter the *Naturalist's Journal* records—

" Brother Thomas, and his daughter, and two sons came. The chaise that brought some of them passed along the King's highway into the village by Newton lane, and down

---

* This word, in these days of universally metalled roads, may be strange to many of the present generation. It means the space between the horse's track and the wheel ruts, which was not worn down.

the N. field hill, both of which have had much labour bestowed on them, and are now very safe. This is the first carriage that ever came in this way."

In December of this year (1784) the great frost, a record of which Gilbert White wrote,* occurred, while his brother Thomas, who was still at Selborne, described it in the 'Gentleman's Magazine,' 1785, vol. liii. pp. 170, 171.

* *Vide* 'The Natural History of Selborne,' Letter LXIII. to Barrington.

CHAPTER VI.

EARLY in 1785 Mr. Richard Yalden, Vicar of Newton Valence, died. His loss must have been severely felt by his friend at Selborne, who refers to it below.

To Miss White.   Selborne, Mar. 9, 1785.

Dear Molly,—We thank you for the tea, which we think very good: and also for the salt-fish, which proves more white, and delicate than usual. Instead of in a parcel, the cod came down in a barrel, which being leaky let the brine out on the kitchen-floor. I therefore told Thomas he should carry it into the cellar. Thomas without much thought took the barrel by the hoops, and was got to the cellar stairs; when off came the hoops, down fell the barrel, out flew the head: in short the stairs from top to bottom became one broken wet scene of barrel-staves, and codfish! Please to send me the price of the fish, and remind your father to charge it to me. Has your father received my Xmass long ann. dividend?

Pray enquire of your father what is the name of the author of the vision of Piers Plowman: he was, I remember, Fellow of Oriel College,* you may find about him in

---

* There is no evidence to support this statement.

Warton's 'History of Poetry.' What says Dugdale of
Canons regular of the order of St Augustine?

While you were with me, and since, I bore up very well
against the severity of the season: but now this return of
winter, with the aggravation of March winds, has quite
over-come me, has given me a feverish cough, and confined
me to the house. The late frosts have destroyed all the
garden-stuff, and, as men tell me, have much injured the
wheat. Everything is very backward: for the ground is so
hard that the farmers cannot plow their stubbles. John
Withal has been able to do some work for some time: but
two of his fingers, one on each hand, are still very sore, and
want much dressing. Mr. Dusuetoy, the curate at Newton,
has just recovered from a purse-proud farmer at Eastmeon
the sum of £200 with costs, having brought an action for
most gross usage, such as calling him, where-ever he met
him, French son of a female dog; spitting in his face at
Church and else where; for having written to the Bishop to
prevent his getting priest's orders, &c., &c., the jury was
special, his attorney Mr. Clement. Mr. Mill, the clergyman
at Faringdon appeared against the smuggler that robbed him
on the Farnham road, and has got him condemned: he is a
most daring fellow, and has twice broke out of gaol, once at
Dorchester, and once at Winton. They have cut all the
beeches on the top of the hanger; have thinned that beautiful fringe, but not destroyed it, so that next summer the
Hanger will be as lovely as ever. I thank your father for
his account of Oaks.* We lament greatly the loss of poor
Mr. Yalden: this is now the second good neighbor that I
have lost in eleven months! Mr. Charles Etty is still at
Gravesend or the Hope, two miles lower: but they are on
the wing, and expecting to sail every hour for India. The
parish sent our sailor-boy, Bridger, up to town, and down

* Published in the form of a letter to the 'Gentleman's Magazine,' 1785,
p. 109.

to Gravesend, hoping Charles Etty would have been able to
have got him a passage to India as a soldier: but after
being at considerable expence for clothing, a watch (the boy
would not go without a watch) and other matters; he could
not get admittance aboard, and is returned to this place.
Some snow lies still on the hill, and there is this day a
great rime on the hanger, but none in the vale with us.
Thermometer 26°.   The ground is as hard as a rock:
Bar. 29·25.

Yʳ loving uncle,

GIL. WHITE.

Mrs. J. White joins in respects.
Crocus's begin to blow.

"Say, what impells, amidst surrounding snow,"* etc.

Writing on March 19th, 1785, Mulso regrets the
death of Mr. Yalden—

"You have lost some very good men and true Christians
from your neighbourhood of late.   For God's sake take care
of yourself and live as long as ever you can to keep up so
precious a character. . . . I fear you are all plunged again
into sadness.   I pray God that the vicinity of your nephew
may produce future scenes of joyousness and happiness."

This nephew, also a nephew of Mr. Yalden's, was
Edmund, son of Benjamin White, who was to be
the new Vicar of Newton.

In April the usual visit to brother Thomas at
S. Lambeth was made, and the *Naturalist's Journal*
records—

"May. 1. Saw one *Swift*, two *house-martins* in Fleet St."

---

* In the second (1802) and subsequent editions of 'The Natural History
of Selborne' these verses were placed at the end of Letter XLI. to Barrington.

Returning to Selborne, a notice occurs of the *Coccus vitis-viniferæ*, with a speculation as to the possibility of its having come from Gibraltar in boxes, etc., received twelve years ago from thence.

"My brother John, in his excellent 'Nat. History of Gibraltar,' which I have by me in MS., gives the following account of this *Coccus vitis viniferæ*," etc.:— *

To Miss White.

Selborne, May 17th, 1785.

Dear Molly,—When I came to look over my melon-seeds, I was surprised to find only one paper of Succades, and those twenty years old: however, I would wish your father would try them; and if Smith was to soak them in water a few hours before they were sown, he might have a better prospect of success.

We had a hot, and dusty journey to Alton, and especially between Cobham and Ripley, where the heaths were like the deserts of Arabia for smother, and fervent heat.

Our grass and corn are in a bad state, and my garden, and grass-plots are burnt up. My wall-nut tree is so killed down, that the foliage will be this year very scanty.

The sycomores have suffered also; and my jasmine is dead. We have cucumbers, and a few asparagus: but Timothy has devoured most of my lettuce. The hanger is beautiful.

As my Barometer is this day below 29·3, we have some hopes of rain; but it seems very unwilling to come.

We return your father and you, and all friends many thanks for all your good offices while we were with you. We have little apple and pear-bloom, and no wall-fruit. Farmer Town is very bad; and Betty Loe is very weak. John Stevens has a son and heir.

* *Vide* 'The Natural History of Selborne,' Letter LIII. to Barrington.

Mrs. Etty expects Lady Young, and Mrs. Rashleigh this afternoon.

Y^r loving uncle,

GIL. WHITE.

My cough is better. My Portugal laurels seem to be dead.

To Miss White.   Selborne, June 25, 1785.

Dear Molly,—I am desirous to address you once more in your maiden state, which, I now understand, is soon to be at an end.

I hope you received my long letter by Edmund, in which I desired to know whether Smith had any success with my old Succade melon-seeds, so as to make them grow : I wished also to have enquired after the cuckow's nest, but, I think, I forgot it.

Mrs. J. White and I have made many enquiries about a maid for you; and one day I rode myself on that errand as far as Kingsley. In consequence of my ride we had a maid offered from that parish, of whom Mrs. J. White has sent you some account by the post. Tell your father I have taken up riding again, and mount my little horse frequently. Inform him that we have also pulled down part of his old cottage, and find it in a very bad state, most of the timbers being worm-eaten beech. In the room of what we have taken down I shall erect a fewel-house; and will endeavour to put him to as little expence as possible. The drought begins to be very serious and stubborn, so that there will be very little grass, and less spring-corn. We water, and water; but our annuals die away." [The rest of this letter is missing.]

On June 29th, 1785, Mary White married her cousin Benjamin, the eldest son of her uncle Benjamin White, and went to live at the "Horace's Head," No. 51, Fleet Street.

The *Naturalist's Journal* at this time records some remarks on the fly-catcher, all of which were not incorporated into their author's book.

"1785, July 25. While the hen fly-catcher sits, the cock feeds her all day long: he also pays attention to the former brood, which he feeds at times.

"Aug. 1 [at Meonstoke]. Fly-catchers in Mr. Mulso's garden, that seem to have a nest of young. The fly-catchers hover over their young to preserve them from the heat of the sun.

"[Aug.] 8, Selborne. Fly-catchers' second brood forsake their nest.

"Sept. 8. Mr. S. Barker came. Planted a *Parnassia* which he brought out of Rutland in full bloom, in a bog at the bottom of Sparrow's hanger."

On September 4th, 1785, John Mulso forwarded a criticism of his friend's lines 'On the Early and Late Blowing of the Vernal and Autumnal Crocus'—

"I received your 'Crocus' in its triple shape, and I like it in all. The original is an ingenious thought, piously as well as poetically imagined, and happily expressed. Of the two translations I like the first and shortest the best, but I do not approve of the *stop* ';' after 'summa potestas'; 'Florarum Deus,' etc. is the answer to the question. If it is added, 'whose power *is* supreme'—'cui summa potestas'—as it must be construed if stopped so strong, then 'ipse' is wanted before 'temperat.' But if it intends 'whose supreme power tempers' etc., then it should have no stop. It is concise and just. The other, more at large, is likewise well done; but the same objection occurs at the same place. And I fear that 'calet' and 'liquet' are applied as *active* verbs, which is not usual Latin, unless I have forgot it."

To Mrs. B. White
   at Horace's head
     Fleet street, London.

                   Sepr. 20, 1785.

Dear Niece,—I fully intended to have made some catchup
for you, and your father; but the mushrooms fail again, as
they did last year, to such a degree that we have not
been able to raise half a pint. Last week Mrs. Burbey
had a respectable farmer, a cousin, with her on a visit.
This person waked in the night, and found himself standing
naked in the cart-way with his night-cap in his hand.
How he came there he could not imagine: but there was
just light enough for him to see that the sash of his
chamber was up; and therefore he concluded he must have
come down from thence. Finding the doors all locked, he
was forced to call up the family, who were astonished to
see their visitor in his shirt, and much confused and
frightened. The poor man received a cut on the sole of
one of his feet, and a small contusion on one knee: and
this was all the harm that he received from descending,
fast asleep, for the space of twelve feet and a half, down
on the bare pavement. Mr. S. Barker is with me; and I
expect uncle Harry this week for two nights; and perhaps
Betsey,* who, I hope, will stay a little. Your father in
law and sisters, &c. are all well. Newton's vicar has got
his harpsichord down; and I can borrow Mrs. Etty's.
Mary and Hannah† are most woefully *overhatted*, so that
they look like *Jenny Diver*, and *Miss Slamerkin* in the
'*Beggar's Opera.*' Tell your father that I have seen no
Serapias in bloom all this autumn.

                   Yr loving uncle,

                      GIL. WHITE.

     * Daughter of Henry White.
     † Daughters of Benjamin White, senior.

After visiting Oxford in October, a large party of relations from Fyfield were received at Selborne.

To Mrs. B. White.   Seleburne, Decr. 1st, 1785.

Dear Molly,—I am very angry with your husband for calling away his wife so suddenly, when in appearance he had acquiesced in her staying 'til the return of her father. I am far from wishing that time may abate his affection for the good woman in the smallest degree: yet I hope to live to see the day when she will be permitted to make me a much longer visit; and to come down and look at her children, which, peradventure may be nursing in this village.

Thomas went this afternoon to the Horse and Jockey to meet Mrs. Etty and Co. who were expected to have been there about three o'clock: but, after waiting till it was dark, returned, bringing word that the ladies were arrived at Alton; but came in so tardy* that they thought it more prudent not to trust our rough roads at so late an hour. I thank you for your letter, which I *expect* by Mrs. Etty to-morrow, but not being able to divine the exact purport of it, I can only foresee by anticipation that you are much obliged for all favours, and *all that*: to which I reply that you were heartily welcome, and *all that*. Hoping, an other year, to find you a more dutiful niece, and less dutiful wife, I remain with true affection   Yr loving uncle,

GIL. WHITE.

On December 7th, 1785, Mulso writes :—

"I heartily join in your satisfaction on the Provost of Oriel's kind promise to your brother Harry's son:† but

---

* Another word which has quite fallen out of general use. Yet Wykehamists keep it in mind, and have substantial reasons for so doing, when *and as* they happen to come into chapel after the bell has stopped. At least it used to be so when Plancus was Consul.

† Henry White's son, Samson, was eventually elected to a Fellowship at Oriel.

will he live to be as good as his word? I hear, poor man, of a dangerous state of health that he is in.

"I am not going to wish you joy of Tortworth living:* I am sure you will never think of it; for if everything went quite smooth there, neither the country, nor the modes of collecting your income are at all to your mind. Get some pretty little sinecure tenable with your Fellowship. Live on at Selborne and be cotemporary with Jack Mulso still. . . . The winter has been so severe that I am glad that you do not trot to Faringdon on a Sunday."

Mr. Churton came to pay his usual Christmas visit at Selborne.

To Mrs. B. White.   Seleburne, Decr. 26, 1785.

Dear Mrs. Mary White,—I thank you for your kind letter respecting raisins, and salt-fish. Concerning the former I shall say nothing just at present: but must desire you to procure me, as soon as conveniently you can, five good Iceland cod-fishes; *two* for Edmund White, and *three* for myself, to be sent down by Findon, the Faringdon Carrier. If your husband has an account with Edm^d please to desire him to charge *two* to him, and *three* to me: or else let me hear, and charge all to me, and I will settle with Edm^d.

You know of course, as well as I, how the matter stands between nephew John,† and Mrs. Kemp. He left proposals behind him, which I much suspect will be accepted. If he does settle among us, I sincerely wish the undertaking may answer his expectations.

We had very mild weather 'til Friday last: on Saturday

---

* In Gloucestershire, an Oriel living.

† "Gibraltar Jack." Mrs. Kemp appears to have been widow of a doctor at Alton, whose practice he was proposing to acquire. The next letter shows that he did settle in the neighbourhood.

morning came snow; since which the frost has been severe. This morning the thermometer abroad at 24°. There are in this village hand-bills sticking up signifying "that on Thursday there will be shot for at the Red Lion at Oakhanger a good fat porker, yards for inches, at a card: all to charge out of the same bag: and also two boar-pigs to be bowled for." Other bills say, "that at Faringdon on the 28th a good watch of one guinea value is to be *played at farmering for*: he that *wins the farm* the three first times to have the watch." Such are the amusements of this neighbourhood at this season of festivity. This evening I had a letter from Mrs. Barker in which she says, that Mrs. Brown is to lie in about the end of March. Upon which I remark, well done Mrs. Brown! well done Mrs. Clement! and so vice versâ: for I think they are well matched, and will run a hard heat. Tell Nephew Tom I thank him for his letter, and will write soon. I can scan his verses and think the measure very pretty.

With due respects I conclude

Your loving uncle,
GIL. WHITE.

To Mrs. B. White.

Jan. 16, 1786.

Dear Molly,—Though I have little to say, yet as James Etty is going to town I would not omit the opportunity of writing by him.

We are to thank you for the salt-fish, which came safe and is very good. Mrs. Clement came to Newton yesterday with her little family, who are to be inoculated at the vicarage at Newton this evening. Nephew Edmund set out for Oxford last Friday, some days sooner than he first intended, and proposes to return by way of London about the 20th of February. The rains and snows are, and have been so great lately that the ground is quite glutted with

wet: if the fall has been as great with you, your two fathers
will be in danger of seeing more liquor in their cellars
than they would wish. Much water has soaked into my
cellar lately. *Dr. White* often calls on us: he has met
with dismal weather in his rounds. Pray ask Mrs. Yalden
where *her* volume of Mr. Churton's sermons is. Did she
bring it down from S. Lambeth to Newton? The reason
of these enquiries is because I have lost *my* book, given
me by the author. There is a copy at Newton, which
I suspect Edmund borrowed of me; but, being a lover,
he does not know whether he did, or no. As a certain
man was looking at his son lately, while he was eating
bread and butter, he remarked, "that he was a sound puppy,
and did not husk while feeding." Mr. Ventris has been
much out of order lately, so that I have not seen him.
I wrote a long letter to Nephew Thomas Holt-White lately,
and hope to hear from him soon. Is Uncle Harry in town?
Boswell's journal* is a comical, pleasant book. I wish he
and Dr Johnson had taken more tours together.

Mrs. J. White joins in respects.   Y^r loving uncle,

G. W.

To Mrs. B. White.

Selborne, Jan. 24, 1786.

Dear Niece,—Though I wrote to you the other day by
James Etty; yet I do not love to let Mrs. J. White's
parcel go off for the coach to-morrow without sending you
a line. To-morrow I am to marry Nanny Hale over the
way to young Farmer Tull of Wick-hill. The young woman
has been bred to habits of industry; and the young man,
I believe, is sober, and steady: so the match, I think,
promises well. Yesterday Mrs. Etty and Co. and nephew
John dined with me: whether I drank too freely among

---

* 'The Journal of a Tour to the Hebrides with Samuel Johnson, LL.D.,
by James Boswell, Esq.,' was published in 1785.

my friends, I cannot say: but in the evening and part
of the night my feet raged and I felt violent pains, so
that I expected the gout at once: but this day, I thank
God, they are better; but still in a grumbling way. Mr.
Powlett, it is said, is going to quit Rotherfield, and to
retire. Is Uncle Harry still in town; or is he returned
back to Fyfield? The quantity of fall, snow and rain,
has been very great this month: already, I think, six inches
and upward. How do the cellars stand affected at S.
Lambeth? Some people continue still to fall with sad
fevers at Alton; some die. The snow of Wednesday,
Jan. 4th was drifted much under the tiles of my roof,
and lodged on the ceilings, where it became very troublesome, and did much damage, when the thaw came.

Y^r loving uncle,
Gil. White.

The Provost of Oriel College, the newspaper says, has got
a daughter.

To Mrs. B. White.   Seleburne, Mar. 25, 1786.

Dear Molly,—Was I to let the bearer to return without
any answer to your late kind letter, I should think that
I had not deserved so kind an attention. Have you heard
of the great good fortune that has befallen Crondall, a
village near Farnham? The people of that parish in a
club, about 36 in number, joined and bought one *quarter*
of a ticket, which came up one of the £20,000 prizes: so
that a number of little farmers, and servants, and labourers
have shared £5,000 between them; and in such a manner
as one would think, would do the gainers no harm, but a
great deal of service. Had that great lump fallen to any
individual amongst them, it would, in all probability, have
driven him quite into a frenzy: but now it is lowered
and diluted into so many parts, there is reason to hope

that all may be made very happy. The prize was not
bought in equal shares; so that many servants and labourers
divided only £35 per man. I wish such a prize, so divided,
would befall my neighbours at Selborne.*

On Tuesday Mrs. Etty and her maidens leave us: and on
Thursday, by permission, Mr. Taylor, our vicar, is to come to
the parsonage house for eight or ten days, and to bring his
bride with him, and to keep his wedding here. They are to
bring, I hear, a man and maid servant with them, but no
lady bridesmaid; and are, I conclude, to be in a very private
way: however it is not unlikely that we shall visit them.
The bride is—"ah! pray, Uncle, tell me who the bride
is." Why, the bride is to be Miss Lisle of Moylscourt near
Ringwood, a lady of one of the best families in the county;
and whose uncle represented this county in parliament
for many years; and whose grandfather wrote the book
on husbandry. I am in a sad fright, having no silk-breeches,
and stockings to make a wedding-visit in. In just such
a fright was Uncle Richard,† when first your mother came
unexpectedly to Newton. John Carpenter has opened a
shop with a great bow-window to the Plestor, in which
he sells ironmongry, hardware, cheese and breeches. Tell
your father that the vast Ash-tree on the Plestor is all
worked into bushels, half-bushels, pecks, gallons, and seed-
laps; and that it was curious to see such a huge, stubborn
mass by art bent and moulded into so many pliable and
shapely implements and utensils. Tell him also that
nephew Edmund is very earnest to procure some rock-work
for the bottom of his shrubbery; and that he, nephew
John, and Ventris are to go all to Bridestone-lane in order
to select out some large rocks for that purpose.

* On one occasion the purchase of a lottery ticket occurs in Gilbert White's
account-books. He had a distinguished example, since John Mulso mentions
that his uncle, Bishop Thomas, had shared a ticket *with his pupil, Prince
Edward.*

† Richard, brother of William Yalden, Mrs. Thomas White's first husband.

I hope you practice every day at your Glass; and that you are by this time perfect mistress of "*Nimini pimini.*"* Inform your father also, that nephew Edm^d has carted *his* flints to the top of the Bostal; and that I hope soon, now the snow is gone, to lay them in that path. Mrs. J. White joins in respects.

I am, dear niece, Your affect. uncle,
GIL. WHITE.

Your aunt seems to be quite recovered from her fall, which was a very dangerous one. Pray present my respects to Miss Rebecca White for her fine present of flower-seeds.

At the end of March, 1786, Mr. and Mrs. Taylor paid a short visit to Selborne. Mrs. B. White's curiosity about the lady was thus satisfied by her uncle :—

To Mrs. B. White.   Selborne [April, 1786].

Your aunt and I think the bride a very agreeable woman, and began to wish they would have stayed longer; but they set off this morning for Blashford. However they seem inclined to come again, when the summer is established. Mrs. Etty, I find by her letter, is a little *jealous* lest they should grow *too* fond of Selborne: but I think her suspicions are groundless; and that the incumbent, now marryed, is less likely than ever to reside, because his lady seems strongly attached to Moyle's Court, the seat of the Lisles, from whence Blashford is distant only one mile. The Lady is tall, and well-shaped, about thirty years of age, and has an easy, engaging address: her complexion is dark, and her hair very black; and no wonder, since her Mother was a Levant woman, perhaps an Egyptian; because her father lived, she says, eleven years at Grand Cairo.

* At this time a miniature portrait of Mrs. B. White, junr., was painted.

The following card was lately put into my hand. " Jⁿ Carpenter, carpenter, cooper, bent-ware maker and iron-monger at Selborne near Alton, takes this method of acquainting the public that he sells all kinds of bent-ware goods, and all sorts of cooperage, chests of drawers, bedsteds and tables, nails, locks, joints," &c., &c.

I proposed to have written by Edmund but he hastened away a day sooner than I was aware.

Y^r loving Uncle,
Gɪʟ. Wʜɪᴛᴇ.

At this time Dr. Chandler was travelling in the South of France, whence he sent his friend at Selborne frequent accounts of the occurrence of swallows, which were duly noted by him in his *Naturalist's Journal.*

To Samuel Barker.   Seleburne, Apr. 17, 1786.

Dear Sir,—Partly through idleness, and partly through infirmity I have too long neglected your late letter. My thanks are due for your curious account of the climate of *Zarizyn,* * and I feel myself the more obliged, because you know I love to study climates. Whether you translate or abridge Dr Pallas, I do not know; but should be glad to see the remaining part of the year, if the subject does not give you too much trouble. I believe all fervid regions afford instances of undulating vapours; that at a distance appear like water. Arabia, I know does; and the phenomenon is finely alluded to in the Koran. In what language does Pallas write?

The summer-like weather of last Friday fetched out *Timothy.* There is somewhat very forlorn and abject in that

---

* This is to be found at pp. 641-646 of the third volume of the original edition of Pallas's 'Travels' (Reise, u.s.w., St. Petersburg: 1776).—A. N.

creature's first appearance after a profound slumber of five
months. When a man first rouses himself from a deep sleep,
he does not look very wise: but nothing can be more squalid
and stupid than our friend, when he first comes crawling out
of his hibernacula: so that some farther lines of Dryden's
ode, (written he supposes on purpose to ridicule tortoises)
may well be applyed to him:—

> "Has rais'd up his head,
> As awak'd from the dead ;
> And amaz'd he stares around."

There was, as I remember, one Abdon, a judge of Israel,
of whom there is nothing memorial, but that he had 40 sons
and 30 nephews. As a father, this chieftain, I must acknowledge, exceeded me much: but as to the matter of *Nepotism*,
I go much beyond him: for I had 42 nephews and nieces
before; and now Mrs. Brown's little daughter makes the
43rd; and I have more at hand, if I do not reckon my
chickens before they are hatched. Nephew John of Alton,
now *Dr. White*, has met with an ugly accident: as he was
descending from his hayloft, the ladder turned, and gave him
a bad fall on the stones; by which he bruised his side, and
dislocated his *left* wrist: but he was not confined one day,
and is getting well. This young man has found employ, and
much riding about: but he must have time to approve himself, before he can expect much prime business. On March
26th Mrs. and Miss Etty left us for some weeks: and on the
30th by permission, came Mr. Taylor, our vicar; and his
bride Miss Lisle of Moyle's court near Ring-wood Hants.
The lady is of a very good family in this county, and niece
to Mr. Lisle of Crooke's-easton; the gentleman who stood
and carried the grand contested election for this county in
1733; but it cost him £10,000. The lady was desirous of
spending part of her honeymoon at her husband's parish.
Charles Etty is expected home in June. Nephew Edm^d, for
which I highly commend him, is parting with all kinds of

farming whatsoever: he lets all his tithes, and all his glebe; reserving only to himself three or four fields for his horses and cows. He will now know what he has to depend on: whereas both his late uncles* were much imposed on; and were subject to all the rabble and hurry of common renters. Edmund I trust, some time hence, will make an excellent neighbour; but has been as yet a very bad one, for his time has been so taken up with various courtships, that he has never been at home yet for ten days together. He marries, I think, in June: but first keeps another term at Oxford. All my apricots were cut off by that violent weather in the beginning of March! So deep was the snow, and so starved the birds, that the poor ring-doves came into our gardens to crop the leaves and sprouts of the cabbages! Hay is become very scarce and dear indeed! My rick is now almost as slender as the waste of a virgin: and it would have been much for the reputation of the two last brides that I have married, had their wastes been as slender. We have just covered the dirty part of the bostal with small flints. The first *swallow* that I heard of was on April 6th, the first nightingale April 13th. The great straddle-bob, Orion, that in the winter seems to bestride my brew-house, is seen now descending of an evening, on one side foremost, behind the hanger. The almanack announces Venus to be an evening star, but I have not seen her yet. Miss Etty is not so well as could be wished: she is low and languid; and often short-breathed. Miss Layton of Alton, Mr. Charles Etty's niece, is lately dead. Mrs. J. White and I thank your mother for her kind letters: the former will write soon.

We think Mrs. Taylor an agreeable woman.

I am, with all due respects,
Your affect. uncle,
GIL. WHITE.

---

* Edmund and Richard Yalden, who respectively succeeded their father, the Rev. Edmund Yalden, in the vicarage of Newton Valence.

From Samuel Barker.

[With an extract.]

Lyndon, July 18th, 1786.

Dear Sir,—I have above completed the history of the
year at Zarizyn, but am afraid that in the names of the
birds you will meet with some confusion, as I have no
assistance but a common French dictionary and the Latin
names are but seldom annexed. I translate from a French
abrigement of the travels of Pallas, Gmelin, and the other
naturalists deputed by the Empress of Russia.

My principal object in writing at present is to inform
you of a change in my situation which is soon to take
place, and which I ought to have been earlier in mentioning
to you. The lady with whom I am to be connected, is of
Northamptonshire, her name is Haggitt, a young woman of
gentle manners and long black eye-lashes, and of course,
you know, everything that is agreeable, &c., &c., &c. I have
had much trouble in finding a house, but believe I shall
at last fix at Whitwell in the house formerly inhabited by
the Isaacs; it is at present much out of repair, but when
put into order, may suit me I think tolerably well. Edmund
White is, it seems, beforehand with me;* present my congratulations to him, and tell him I shall follow his
example as fast as I can.—I have been last June a long
and delightful tour in the North, with my uncle Henry, the
extent of which was Northward as far as the mountain
Skiddaw, and westward to Moel-Enllyn, a mountain near
Ruthin in the vale of Clwyd; but I have enlarged so much
on the travels of Professor Pallas, that I have no room left
for my own, and must conclude with respects to all friends,

Your affectionate nephew,

SAMUEL BARKER.

* He was married on June 20th to Miss Anne Blunt, as his uncle recorded in his *Naturalist's Journal*, adding that this increased the number of
nephews and nieces to forty-five.

To Samuel Barker.
Seleburne, Augst. 1, 1786.

Dear Sir,—As you know I am fond of the history of various countries, and in particular love to study and compare climates; it was very kind of you to take so much pains to compleat the history of Zarizyn for a year.

I return you my thanks for your making me your confidant in a matter of so much moment as that of your taking a wife. You, no doubt, will make a prudent choice; and then there will be a good prospect of your being happy in a state where both parties must concur to render the change agreeable. As it is much the fashion now for the man and his wife to set out on a visit as soon as the ceremony is over, we shall be glad to see the lady and you here, where our new niece will meet all proper respect, and every attention from myself and Mrs. J. White. Edmund's wife made my nephews and nieces 45 : and we expect every day to hear that Mrs. Ben. White has added one more to the number : so that according to appearances the lady we are talking of will be the 47th.

We have experienced a very dry and hot summer; most part of June was sultry : yet we had a good crop of hay; and have a fine prospect for wheat, which is very tall and even : the hops also look well : but of late the pastures and meadows burn, and the gardens suffer greatly. My grapes are very forward, and the crop large. Plums we have none, and no wasps yet.

When I see you, you must tell me all the circumstances of your long tour, which cannot fail to entertain. I only fear that after your eyes have been stretched with the sight of Skiddaw, &c., that you will despise the mole-hills of this district, which once used to delight you so much. My intended niece, I trust, will be pleased with our hangers and prospects. Whitwell, I think, is a pretty situation. In the year 1742 I spent a very pleasant long vacation there.

Tell your mother that on the 10th of this month she and I shall have a new sister.* Verses have been written on ladies *eye-brows*; but you talk of the beauty of your mistress's *eye-lashes*: in that matter as far as I remember, you speak like a Turk. Now you talk of ladies, can you repeat "Pretty, pretty Peggy Haggit" three times in a breath? We expect Mrs. Etty from Beaconsfield every day. Her son Charles, it is to be hoped, will soon return from Bombay.

Your loving uncle,

GIL. WHITE.

Little Tom Clement is visiting at Petersfield, where he plays much at Cricket: Tom bats, his grand-mother bowls, and his great grand-mother watches out!!

To Mary Barker.   Selborne, Octr. 25, 1786.

Dear Niece,—I received your favour of Octr. 12th, and rejoice to hear that my nephew Mr. Barker has made so prudent a choice, and has so fair a prospect of happiness in the matrimonial state. He is to live, I find, at the parsonage house at Whitwell, where I spent three very agreeable months as long ago as the year 1742, when I was a very young man.

Present my respects to your father, and tell him that the caterpillars of *phalænæ* devoured all the foliage of our oaks in the bud, and therefore of course there could be no acorns: but that the beeches were loaded with mast; and that I was not unmindful of his injunctions; but have employed people to pick up a quantity of seeds from those trees, which I intend shall be cast into the bushes on the down. We had a wet, cold August and September after a dry spring, and hot summer. We have grapes in vast abundance, that

---

* On August 20th, 1786, Benjamin White, senr., married (secondly) Mary, widow of the Rev. Richard Yalden, Vicar of Newton Valence. This lady appears in the folding north-east view of Selborne, which forms the frontispiece to 'The Natural History of Selborne.'

were very forward in July; but they are not so delicately ripened as in some more favourable autumns, though now good. The beginning of this month deluged all the country, and had like to have blown us all away : the tempests and torrents were dreadful! From the 4th to the 11th of this month inclusive the quantity of rain was 5·04! but now we have delicate weather, and a fine wheat season. The late election at Salisbury has done my Nephew John much honour: but neither he nor his mother are elated on the occasion, because he quits a little certain business in hopes of greater. He certainly was getting ground at Alton. Should he succeed at Sarum, there will be more field-room for getting money than in our poor rough district; and so there had need : for the Infirmary brings neither salary, nor emolument, but only credit, from the supposition that the surgeon is a man of skill and merit in his profession.

Brother Thomas is here; and brother and sister Benjamin and Mary at Newton: they join in respects. I am glad to hear that Mr. and Mrs. Brown have left Uppingham.

I am, with all due respects,
Your affectionate Uncle,
GIL. WHITE.

Our hop-planters returned from Wey-hill fair with chearful faces, and full purses, having sold a large crop of hops for a good price. The reason was, because the Kentish hops, which were a fortnight behind, were blown away by the tempests. The parish of Selborne will be much benefited by the hop-plantations, to the amount, some say, of near £2,000. The women had a fine picking, and earned 2s. 6d. per day. Uncle Harry has built him an hermitage at Fyfield, on which Samson White has written a good copy of verses. Mr. Twopeny is just married.

From the *Naturalist's Journal*—

" 1786. Sep. 23, Brother Thomas and sons came."

The following whimsical experiment is here recorded.   These articles—salt of tartar, oil of vitriol,
a tea-kettle, sea salt, wood ashes, coal ashes, nitre—
were placed in concentric circles.

"With the above mentioned articles Bro^r Thos. has attempted to make a fairey-ring, circle within circle : and we
are to take notice in the spring which circle, and whether
any, will produce grass of a deeper green than before.   The
tea kettle . . . was set out, time after time, full of boiling
water.   The circles made with oil of vitriol, with sea salt,
and with salt petre have discoloured the grass : those with
Sal Tartar, wood, and coal ashes have no visible effect at
present.   The grass seems killed where the tea-kettle stood.

" Novr. 22. I sent a woman up the hill with a peck
of beech-mast, which she tells me she has scattered all
round the down amid the bushes and brakes, where there
were no beeches before.   I also ordered Thomas to sow
beech-mast in the hedges all round Baker's hill."

Later in the year a visit was paid to Fyfield.
After the return to Selborne Mr. Churton came as
usual at Christmas time.

CHAPTER VII.

To Thomas Barker.

Selborne, Jan. 10, 1787.

Dear Sir,—I have herewith sent you the Selborne rain, an account of which, I think, has been kept very exactly: but know nothing of the Fyfield and S. Lambeth rain. There fell such a glut of rain in the beginning of October that men were in some pain about the wheat season: however, such lovely weather followed quite into November that the sowing time was unusually good. Again during the 14 first days of December there fell 5 inches of rain: this deluge washed our malm-grounds sadly.

As to strong beer at Mr. Yalden's, I can say nothing about the management of it, because John Pullinger, who had the sole conducting of it, has left Edmund White: I only know that my strong beer is much admired by those that love pale beer, made of malt that is dryed with billet. My method is to make it very *strong*, and to hop very *moderately* at *first*; and then to put in it, at two or three times, half a pound at a time of *scalded* hops, before I tap it. This is the Wilts method, and makes the beer as fine as rock-water. As my family is small, I never brew more than *half a hogshead* at a time; but then I put malt at the rate of 13 bushels to the *hogshead*, and only 3 pounds and a half of hops at a brewing. I tap my half-hogsheads at about 12 months old; and always brew with rain water, when

I can. The tank at Newton is made of brick: their beer was, and is, often good; but their water, when drank by itself, has a filthy taste of lime and moss. Their table beer does not keep in summer.*

Please to present my best thanks to my sister for her kind charity which will be very acceptable to our numerous poor. Mrs. Etty is here, but will leave us soon, perhaps 'til midsummer.

Y^r affectionate servant,

G. WHITE.

The crop of beech-mast was prodigious, and of great service to men's hogs, which were half fat before they were shut up. Between mast and potatoes poor men killed very large hogs at little expense. Tom Berriman's hog weighed 16 scores; yet eat only seven bushels of barley-meal: whereas without the help above mentioned, he would have required 20 bushels.

Dame Berriman is much disordered in her mind, and very violent. I sent a woman to scatter some beech-seed in every bush on the down.

Mrs. J. White joins in respects. Barometer has been very high for some days; on Monday it was 30·3.

From the *Naturalist's Journal*—

" March 27. Swallows were first seen this year at Messina in Sicily.

" April 6. Nightingal sings at Citraro in the nearer Calabria.†

" [April] 13. Sam[son] White elected a fellow of Oriel College in Oxford."

---

* Amongst Gilbert White's papers there is " An account of the brewings of strong-beer. A chronicle of strong-beer, and raisin wine," beginning March 24th, 1772, and containing regular entries of the brewing and tapping of strong beer and raisin wine, and bottling out port wine. The last entry occurs thirteen days before his death—" June 13th, 1793. Tapped the other barrel of raisin wine: it is well-flavoured."

† Communicated, no doubt, by Dr. Chandler.

His nephew's success is in itself evidence that Gilbert White was anything but a "persona ingrata" at Oriel.

On May 6th, 1787, Mulso writes to his friend upon the subject of a subscription to the new library then proposed to be built by Oriel College, for the books bequeathed by Lord Leigh. He continues—

"I sincerely wish well to the Society; it furnished me with a friend whom I continue to value, and shall look upon as one of my blessings to the end of my life. You know less of his worth than other people, so I shall not put you to the trouble of guessing at him. . . . I heartily wish you joy of having got your brother Harry's son into that Society, as I know you had set your heart upon it. It is indeed a fine provision for a young man. I hope you did not signify a willingness to resign your own, in order to facilitate his Fellowship. Keep that eligible Bisk* in your sleeve and cease to curatize; it is too great a trouble for you: 'solve senescentem.' You may do what duty you please, but do not be under the necessity of doing it, or the solicitude of getting your place supplied. This is my serious advice, and the wish of all who love you."

The advice of his easygoing friend did not commend itself to the Curate of Selborne; who, with more wisdom, had neither the wish nor the intention to live an idle life.

In May, 1787, the following note was made in the

* Bisk or bisque, a term in the game of tennis for the odds one player gives the other. Mulso's expression, signifying "to have an advantage to fall back upon," seems to have been proverbial, as the 'New English Dictionary' cites two examples of its use; one in 1713—"He (like a compleat Politician) reserves always a Bisk in his sleeve (a Phrase we Tennis-players use)."

Naturalist's Journal at South Lambeth, where its author was visiting, and, as usual, using his eyes :—

"May 21. A pair of red-backed Butcher-birds, *Lanius collurio*, have got a nest in Bro. Tho. outlet. They have built in a quick-set hedge.

"In outlets about town, where mosses, lichens, gossamer, etc. are wanting, birds do not make nests so peculiar each to its species. Thus the nest of the chaffinch has not that elegant appearance, nor is it so beautifully studded with lichens as those in the country; and the wren is obliged to construct his nest with straws and dry grasses, that do not give it that roundness and compactness so remarkable in the edifices of that little architect.

"June 2. Hay is making at Vauxhall."

The following interesting letter from Mr. Churton was written at Selborne in the absence of the Naturalist, who had gone to London on the day previous. It seems unnecessary to print all the other letters from this correspondent to Gilbert White in the present writer's possession, especially as Mr. Bell has done so; but by the kindness of a descendant, two letters to him from his Selborne correspondent, in addition to those printed by Mr. Bell, are now added :—

From the Rev. R. Churton.
Selborne, June 6, 1787.

Dear Sir,—I am just arrived from Waverley, and very sorry not to find the master of this hospitable mansion at home. I did not know that I should be at Waverley these holidays till just before I set out thither; and when my plan was fixt I purposed at several times to write to enquire whether you were at Selborne; but one or other avocation prevented me. So here I am; and your bread and butter, and cream and tea and sugar, will shortly suffer great depredations. However, in some respects I hope you will be the better, aye, and the richer, for my visit. In the first place I bring you an Anglesey penny from the fair hands of Miss Loveday; who, I hope, by this time is in perfect health. When I called at Caversham on Whit Tuesday a bad fever was just gone off, but she still kept her bed. Of her friends, however, she was not unmindful, and she sent me down this coin with a commission to bring it hither. I never saw Mr. L. in better health or spirits, though his leg, which he bruised some time ago and neglected, is not well, as it would be soon if he would rest it before him; but he prefers a wounded leg with activity to sound limbs and idleness. This incomparable friend of ours, who knows everything, presently showed me the 'Annals of Waverley' in print, among some other tracts published by Gale. Dr Adee, M.D., whom you knew probably, collected a history of Waverley Abbey; and my friend Dr Bostock has a transcript of it. He has made considerable use of the annals, and appears to have put together all, or nearly all, that is to be met with on the subject. I left a paper for you at Fleet Street, which said that the heart of Peter de Rupibus was buried at Waverley, and his body at Winchester. The Hist. of Waverley mentions this; and Dr Adee adds "that when Mr. Child first came to the place, a heart was dug up in a leaden pot, and preserved in some liquor." Simon de Montfort is also mentioned; but this, I think, I extracted on the said paper.

No Mr. White, no Mrs. J. White, no Mr. Edmund White, no Mrs. Etty! Alas poor Selborne! thy grotesque lanes, thy romantic vales, thy delightful walks, thy verdant hills, thy extensive prospects deserve to be honoured by other inhabitants than the philosophic Timothy in the beginning

of June! Here, however (for I have almost done mischief
enough to the loaf), here "Let me wander all unseen, By
hedgerow elms and hillocks green," in fields somewhat more
fertile than the Surrey hills, where the largest of the trees
first planted by O. Hunter is about 3 feet in girt, after growing, I believe, more years than I have been growing; but
then in height they have far outstript me, to say nothing of
my friend the Archon of Rolle,* who honoured me with a
letter yesterday after a half year's silence. He says not a
syllable about returning to England; but if he has left
Rolle, as perhaps he may before a letter arrives, it will be
forwarded. He says the English literature and nation enjoy
in Switzerland a degree of esteem which is very flattering to
a lover of his country, and that it is surprising to see the
number of English authors to be met with in the libraries of
gentlemen in the delicious little town where he was when he
wrote to me.

I inclose you a letter from the Wanderer (Thicknesse
the traveller); how *instructive* it may prove I know not.
Mr. Burby tells me he saw a letter from C. Etty, which was
forwarded to Mrs. Etty, and that he apprehends he is on the
English coast, if he is not landed. I was much indebted
to the hospitality and conversation of S. Lambeth during my
visit to the metropolis at Easter, which was not so long as it
would have been if the smoke had not given me a wretched
cough, which the air of Oxford and the country removed
some time ago. I am afraid I shall not see Selborne again
this summer, as I am bound for Cheshire towards the end of
the term, which begins to-day. I came across the country
from Waverley by the Holte, through Kingsley, and along
the edge of Wolmer, and never was much out of my way I
believe. Some of the hills hereabout I knew as I approached them; but there was a clump of trees on a

* Dr. Chandler.

promontory to the left of the Temple nearer Empshot,
which disturbed me a good deal. I thought I must have
seen and remembered such a prominent feature (if you allow
fashionable expressions) in the landscape. I am much
obliged to you for the kind letter which I found in my room
on my return to college after Easter. And now let me
inquire after friend Timothy. He looks very well, and says
not a syllable of a late elopement. Perhaps he is ashamed
of it; and yet who knows whether he was not going in
quest of his master, and if he had not speedily been brought
back he might possibly have surprised you by an unexpected
visit at S. Lambeth. Thomas tells me that C. Etty *is*
arrived in England, which I am very glad to hear. I saw
Mrs. Etty for five minutes at Beaconsfield, on my way to
London. The rain, which is just set in, will, I hope, be
of service to the country; but I could gladly have excused
it for three hours longer, one to walk about here, and two
to ride back to Waverley. My great coat I very wisely left
at Reading. I *might* make *that* in my way to Waverley;
but then I should run a risque of losing my dinner, which,
at a proper interval after breakfast, is an object of some
importance. It still rains, and I am still, dear Sir,

Your most obedient and much obliged servant,

R. CHURTON.

The *Naturalist's Journal* at this time records an
observation and suggestion which show the practical
bent, as well as acuteness of its author's mind.

" July 8.   Hops are diecious plants: hence, perhaps it
might be proper, though not practiced, to leave purposely
some male plants in every garden, that their farina might
impregnate the blossoms. The female plants without their
male attendants are not in their natural state; hence we may

suppose the frequent failure of crops so incident to hop-grounds; no other growth cultivated by man has such frequent and general failures as hops."

This note was not used in 'The Natural History of Selborne' as published by its author, though it appeared in the selection from his MSS. made by Dr. Aikin, published in 1795. It has now long been the practice to plant male hops in hop-gardens.

On July 23rd, 1787, Mulso recurs to the old subject of his friend's deferred publication.

"Did you consult your brother about your book, and its publication? I feel impatient. As it is your only child, I hope you will not let it be a posthumous one. You cannot imagine the pleasure you would take in daddling and nursing it, and in the speeches that would be made you on its being so promising, and the features of it so handsome. Then the pride you would take in seeing it dressed in its red and gold, and keeping company with Lord Leigh in the new library at Oriel. But seriously speaking, your diffidence prevents a great deal of credit to yourself, and of satisfaction to the world. In point of profit there is certainly a *white* day to every author; which, if you seize it, is well; if you let go, it is difficult to recover. The aid of your brother in giving a *ton* and a currency is of vast importance; and the zeal of your friends to recommend it and forward its notoriety. All this depends on the present time; and will grow languid and cold, when you are less on the public stage yourself, and cannot second the efforts of your friends. Too frigid caution will make you listen to discouragements; and, believe me, there is more jealousy stirring in the world

than you are aware of. Be bold therefore, and come forth;
' Sume superbiam quæsitam meritis.' "

On Sept. 6, 1787, he writes again :—

" You delight me with the account of your being in the
press. I have written to my brother [T.] Mulso to bespeak
a set of the first impression of your brother Benjn, and I
hope you will second me in it, that what I have of yours
may receive no disgrace after it leaves your hands."

From the *Naturalist's Journal*—

"Fyfield, Nov. 8. The Fyfield comedians performed 'Much
ado about nothing,' with the Romp.
" Nov. 13. The Fyfield players performed ' Richard the
third.' "

These versatile comedians were the children of
Harry White and their friends.

In the times now treated of printing off a book
must have been a long business, since the following
letter was written about a year before the publication :—

To Mrs. B. White.   Seleburne, Nov. 26, 1787.

Dear Molly,—Pray give the opposite letter* to your
husband, and desire him to insert it among those addressed
to Mr. Barrington, but before the three or four last which
all concern the weather at Selborne. This will soon be
followed by an other, which I shall also extract from my
Journals. My thanks are due for your late kind and

* ' The Natural History of Selborne,' Letter LX. to Barrington.

intelligent letter, and quotation, and for your present, and all other good offices. I paid Nurse Abor up to Michaelmas last, as long ago as the first of October. We suppose 'The Wealth of Nations,' which was left behind in the hamper, went down to Fyfield in Mrs. J. White's trunk. Sam[son] White is a very fortunate young man : for now the Provost and Fellows of Oriel Coll. have elected him to a good Exhibition which is to continue three years, viz. 'til he has taken his Master's Degree.* This benefaction was given us, and two more of the same value, by Dr Robinson, Bishop of London, and first Plenipotentiary in the reign of Queen Anne at the making of the peace of Utrecht : he also built us a wing to our College.

But that I may not be wanting in the most momentous part of my Natural History permit me to add, that your son is perfectly well and jolly, and much disposed to eat my roasted apples ; and promises another double tooth : he promises also to be a Connoisseur, for he takes much notice of the engravings on my wainscot, which appear to him to be different from Dad's naked walls. Thanks to your father for his letter, which I will answer soon.

Y^r loving uncle,
GIL. WHITE.

I procured Sam's exhibition for Uncle Harry between 30 and 40 years ago.

Pray let this letter stand *the last, before* the letters to Mr. Barrington describing the weather of Selborne, in number, I think, *four*. Many thanks for your careful corrections of the proofs, which are very exact. Where you find a word not marked thus ——, print it in *Italics* when you think it expedient. Your father-in-law, by his In-structions to Mazel has much improved the front of the

---

* A further instance of the goodwill of the College to Gilbert White.

Vicarage. Desire your father to attend to the *barometer part* of this letter, and to correct errors. I fear I have been negligent in not marking several words thus —— as for *Italics*; and especially the names of birds in letters 39 and 40 to Mr. Pennant.

Manakin is very jolly! Mrs. J. White will write next post.

GILBERT WHITE'S BAROMETER.

CHAPTER VIII.

On New Year's Day, 1788, Bishop Brownlow North issued from Farnham Castle a paper of questions to each incumbent in his diocese, to be returned before the last day of March in the same year, the object of the inquiry being to gain information prior to an approaching Episcopal Visitation.

One of these papers of questions found its way to Selborne. The return was made by the Curate in charge there in his usual careful manner, and is now in the muniment-room at Farnham Castle. It is headed, "Answers to the several questions respecting the parish of Selborne," and contains particulars respecting the extent of the parish, its number of tenements, population, baptisms, burials, and marriages. It continues—

"The Curate for the present (who is not licenced) is Gilbert White, A.M., nominated by the Vicar, the Rev. Christopher Taylor, B.D. For more than a century past there does not appear to have been one Papist, or any Protestant dissenter of any denomination.

"Selborne is not able to maintain a schoolmaster; here are only two or three dames, who pick up a small pittance by teaching little children to read knit and sew."

The present writer happens to possess some of the receipts given to Gilbert White by these Selborne dames, when he paid them for their services out of his grandfather's bequest for that purpose. Usually the form of receipt was written by him; when this was not the case the spelling and handwriting of the ladies leave a good deal to be desired.

The return concludes with an account of the charitable institutions of the parish. It was

"Given in to the Lord Bishop of Winchester, by Gil. White, Curate of Selborne, March 25th, 1788."

To Samuel Barker.   Seleburne, Jan. 8, 1788.

Dear Sir,—It is to be hoped that you are not so punctual a man as to register all the letters that you write to your friends; because the distant date of your last epistle to me would reproach me with neglect and negligence towards one of my near relations. I have been very busy of late; and have at length put my last hand to my Nat[ural] Hist[ory] and Antiquities of this parish. However I am still employed in making an Index; an occupation full as entertaining as that of darning of stockings, though by no means so advantageous to society. My work will be well got up, with a good type, and on good paper; and will be embellished with several engravings. It has been in the press some time; and is to come out in the spring. It pleases me much to find that you still pursue your botany. I had reason to suspect that your noble neighbour had a propensity to the same enquiries, because I have sometimes met him at Curtis's garden. Brother Thomas thinks it may be best to cover the Ginkgo* a little in severe weather. We have had a very deep snow, which began on Sunday Decr. 23rd and lasted for two or three nights, and days, so that several of our hollow lanes became impassable. The turnpike through your village must be a very pleasant circumstance, and prevent such inconveniences; to which, I remember in old days, it was very liable. I recollect to have heard Mr. Isaac say, that they had often been snowed up; and that he had shot woodcocks and snipes from his best chamber-window as they came to feed at the fine perennial spring from whence your parish† takes its name. Mr. Charles Etty left us last Friday, and went to his ship, the Duke of Montrose, now lying at Gravesend, in which he is soon to sail for Madras, and China, as third mate. The wicked woodcutters entered our Hanger this day, for the second time, in order to fell some more of our beautiful beeches. Last year they cleared as far as the *shop slidder*; and will strip now as far as *Hercules*! If my niece does not come, and see the remains of that sweet pendulous covert next summer, she will scarce be able to conceive how lovely and romantic it once had been. Sam White is a very fortunate lad: for not long since the provost and fellows of Oriel elected him to a good exhibition founded by Dr Robinson Bishop of London, which he is to enjoy for three years. I have just sent your father an account of the Selborne rain during last year: it will again greatly exceed that of Rutland.

S. White has undertaken to translate the Prognostics of the Greek Poet *Aratus* into English verses. It has never, it seems, been rendered into our Language: but was so

---

* The ginkgo is the pretty Japanese tree, with leaves like the maiden-hair fern, *Salisburia adiantifolia.*—A. N.

† Whitwell in Rutland.

admired by the Romans, that Cicero and others thought it worth their pains to give a Latin version of it for the amusement of their country-men. Virgil, I fear, that notorious poacher of everything that was elegant in the Greek tongue, has gleaned up every fine Image, and transplanted them into his Georgics. It is remarkable enough that there is now sitting at my elbow an Oxford gentleman* who is deeply employed in making an *Index* also: so that my old parlor is become quite an *Index manufactory*. Mrs. J. White joins in best respects, and the good wishes of the season, to you and lady: and I am,

with all due affection,
y^r loving uncle,
GIL. WHITE.

To Benjamin White, junr.

Seleburne, Feb. [1788].

Dear Sir,—I received your letter concerning Mr. Pegge, but cannot, as you well know, promise any thing. At the same time also I received, by Molly, six more sheets of clean proofs, so well corrected, that I have not met with one error. And indeed the *errata* are so few, that at present they will go into a small Compass, and are as follow:

> p. 11, l. 13. for *scences* read *scenes.*
> p. 31, l. 15. for *teems* read *teams.*
> p. 91, l. 7. d. comma, and for *or* read *of.*
> p. 219, l. 15. for *no tbe* read *not be.*

As I find you advance apace, I have by bearer sent up my Antiquities, because I do not find myself able to correct or improve them any farther. You will be pleased not to be offended at the vague spellings of the names of *men,* and *places,* but to take them as you find them in their places, because centuries ago men had no criterion to go by, but

---

* The Rev. R. Churton.

spelt just as it happened, even their own names often not twice alike. Should not the quotations from the documents be printed in *Italics*? You will, I conclude, have a title-page to the Antiquities, to which the priory-seal will make a proper vignette. The great N. view of Selborne is engraving, I understand, and will be opposite the first title-page: the view of the Hermitage will then best perhaps appear opposite p. 62: in which mention is made of it. That the documents may be kept safe together, I have numbered them, and put them in a paper bag.

As fast as I receive my proofs, I continue to enlarge my Index. The title-page to the Nat[ural] Hist[ory] is furnished with apt Motto's. My thanks are due for all your good offices; and for your late trouble in purchasing me £200 stock.

Y^r loving uncle,

GIL. WHITE.

I take the liberty to return the book of Royal Forests, &c., from which I have extracted some information. You will also receive my preface, or advertisement. Concerning the disposition of the Hermitage print please to consult your father, and father in law. Might not the Hermitage print come in well at the back of the first title-page, or as a tail-piece to the Natural History?

Please to observe that all æ diphthongs, as *musæ, phalænæ*, &c. are always written—*muse, phalene*, &c.—in old records.

Where there is any very bad Latin in the evidences, or strange uncouth spelling, please to put in the margin—*sic*.

None of the *errata* pointed out in this letter was corrected in the first edition of the book, and consequently it would seem that it was already at least partly printed off at this date. The first three of them, with others, were printed on an inserted slip.

To Mrs. B. White.

Selborne, Feb. 3rd, 1788.

Dear Molly,—Tho' I wrote to your husband but yesterday; Yet I don't love to omit writing again by your father, who is to go from us to you to-morrow.

Thanks for your letter by Farmer Keene, and the clean proofs by the same hand. Tell your husband that I wish to have the Gentleman's Magazines remain in London for the reason given, because the wanting may be supplyed perhaps some time hence.

In your father's portmanteau I send up my camblet surtout, and desire you would please to order Bunce to bring you some patterns out of Hand, and to make me such another, and *full as big every way*; cut high behind, but with a turn down collar. Pray charge him to make it *full* as *big* and full as *long*. I send you two pheasants, which were killed on my farm at East Harting, and sent me by E. Woods, and G. Hounsom: one of them I intend for yourself, and one for my sister Ben.

You may depend on it that we shall be glad to see you, when you are disposed to come. The *learned pig** is very well, and very intelligent. "Moreover his mother made him a little coat, and brought it to him from year to year, when she came up with her husband . . ." 1 Sam. ii. 19. The text above I leave you to apply as you think proper: I shall only remark, that it proves affectionate mothers to have been the same in all ages.

Yr loving uncle,

GIL. WHITE.

Pray bring down both my Surtouts, and as many sorts of seeds as Miss Rebecca has to spare; but not many of a sort: and at the same time *more clean proofs*.

---

* His niece's child Benjamin, at nurse at Selborne; "Manakin" in a former letter.

To Mrs. B. White.   Feb. 26, 1788.

Dear Niece,—We were glad that you had a safe, and not disagreeable journey to town. The barometer on Thursday last stood at Selborne at 28·3—and at Newton at the same time at 28 ! and yet we had no drowning rains at that period, nor stormy winds ; but much dripping rain since.

Mrs. Edmund White and her son *Richard Yalden* White go on well, and are in a good way. We think Edm^d has done well in showing respect to his benefactor.

I have been thinking that I have nothing to do with *fox-hunting* parsons ; they must do as they like best :—and therefore be pleased to erase my reflection on them, which you will find among the notes to my letter respecting the *Notabilis Visitatio.** We now recollect, with regret, that we never gave you, nor your father any of his *Vidonia* wine !† The day you left us Manakin's surprize was very great in the afternoon, when instead of his mother, and grandfather he found the parlor full of strangers. He surveyed Mrs. Clement often from head to foot ; and was astonished at the tuft of ostriches feathers, which nodded on the crown of her riding-hat. Mrs. C. brought Jane, and Martha, and her Nursemaid *Zebra White sic.* Mrs. Chase, or Miss Greene must, I think be mistaken about Mr. Churton's Doctorate. My gout is rather better ; and so is that of Mrs. J. White but I think her spirits not good at present. We have very soft, mild weather at present, and a rising barometer.

Y^r loving uncle,

GIL. WHITE.

Richard Yalden White encreases my nephews and nieces to the number of 51.

If the engraving of the great N.E. view of Selborne

---

* It was, however, not erased. *Vide* note to Item 11, Letter XIV. of ' The Antiquities of Selborne.'

† A Canary wine.

village should be finished before the Natural History is all printed, I should be glad if your husband could send me one down with the clean proofs.

I have got duplicates of sheet F. viz. from p. 33 to p. 40 inclusive: the last sheet that you brought me is G g. p. 225. If any objection or difficulty arises in the MS. let your father [*i.e.* Thomas White] decide the matter.

On a blank page in this letter Benjamin White, junr. made a note—

" It will not be necessary to print the quotations from the Documents in Italics.

" The Priory seal will make a very good Vignette to the Antiquities, and The Hermitage to the general Title; but with your leave the 'where the Hermit hangs,' etc., shall be erased.

"The great view of Selborne is not yet finished; great part is done. B. W. will observe the rest of the directions."

To Mrs. B. White.   Selborne, Mar. 13, 1788.

Dear Molly,—My thanks are due for six sheets of letterpress, clean proofs, which I received by means of Mr. Webb: the last sheet of which is N n. Two sheets more, I conclude, will comprize all the Natural History. In return for your care about my brat, I have the pleasure to inform you that your boy is perfectly well, and brisk: and that his nurse is better. On the other side of the paper you will find the epitaphs of your great great grandfather Sir *Samson White*, and your great great uncles, *Henry White*, Alderman, of Oxford; and of *Francis White*, Fellow of Baliol Coll. in the same University. These inscriptions were lately copied,* and sent me by Sam White; and may prove some entertainment to your father and father-in-law,

---

* From their monuments in St. Mary's, Oxford.

as they were to me, who though I have often seen them formerly, yet had forgotten many circumstances. Your husband also will be pleased to see some account of his ancestors. This fierce weather appears very formidable to me, who, though the gout hangs still about me, shall be obliged, if at all able, to set out for Oxford on Easter Monday. Please to send Mrs. J. White word whether Mrs. White of L[ambeth] chuses to have Mrs. Etty's dark tortoise-shell cat. Mrs. E. talks of leaving Selborne in the Easter week.

Y^{rs} &c.,

GIL. WHITE.

To Mrs. B. White.

Friday, Mar. 21, 1788.

Dear Molly,—I have just this minute received your kind letter of the 20th. As I have always looked upon your father's interposition as a singular advantage to my work while printing off, I shall most readily desire that his correction may take place: and any other that may occur in the course of this business. As to the Antiq. letters being addressed to some body, I cannot well tell what to say: though I think if to any, it should be to D^r. Chandler; because he has taken so much pains, and has contributed so much to make them what they are. But as the D^r is too far off to be consulted, and may perhaps be displeased should we take any such step without his approbation, I think it may be best to let the letters go as they are, addressed to *no one.* Your husband is the best judge of the expediency of printing in *Italics* or not; I meant with respect to those documents applyed in the body of the work, not those in the appendix. The Priory-seal will appear well, I think, in the title-page to the Antiquities. If there is any objection to the line under the Hermitage, let it be erased. In the last parcel of clean proofs at 237 p. line 2 there is the following error, which injures the

sense of the motto:—instead of *ast* aves solæ vario meatu,
it stands—*est* aves solæ &c.*

Mrs. J. White desires me to tell you that your boy is
perfectly well: and that she will send you particulars soon.
We were much surprized last night to hear that Mr.
Charles Etty was at the parsonage: he has been lying for
weeks at the Motherbank, near Spithead; but is now to
sail with troops, he supposes, next week. He looks well.
Excuse haste; as I want to send this to Alton this day.

Y^r loving uncle,
GIL. WHITE.

Mrs. White thanks you for your letter to her.

A fragment of a letter to Mr. Churton, written
in the spring of this year (1788), contains the
following interesting reference to the old heresy of
hibernation :—

"Pray go to Waverley: you will take the *Hirundines* just
in the nick of time, just as they are stretching and yawning
and rubbing their eyes; and struggling to loose themselves
from the chains of torpidity with which they have so long
been bound. Only one naturalist that I know of ever dug
out the holes belonging to bank-martins: he made no discoveries—yet you may: and may hereafter figure in the
memoirs of the Royal Society."

The annual Easter visit to Oxford was made at
the end of March. In June the occurrence of the
red-backed butcher-bird, or "flusher," in Benjamin
White's outlet at South Lambeth, was noted in the
Naturalist's Journal, which also records an account

* Corrected in the *errata* slip.

of the crop of lucerne grown in his brother's field, and the exact time which it lasted for his horses' feed.

On July 21st, 1788, Mulso writes :—

"I rejoice excessively at you being now committed to the press. As to your frights and fears, they become you well enough as a modest man, but they are unnecessary as an author. But I will put you in great heart. D^r. Chelsum told me that he had *seen your book*, that it seemed a very promising performance, and likely to get into great favour; that it was well put forth and decorated with very pleasing prints and views. Dr Chelsum is a man of knowledge, a Connoscenti, and deep in Virtù. What would you have more in your pre-existing state? This is but your embryo glory; your material and substantial happiness and enjoyment is to come."

A family tradition has descended to the present writer that Gilbert White feared that the public would "laugh at an old country parson's book," but that his scruples were overcome by the advice and exhortations of his brother Thomas, who promised to himself review it in the 'Gentleman's Magazine,' which he did. That its author had, or at least affected, modest views as to his book's success is shown by some amusing lines which he wrote in January, 1789, in which he predicted a very humble fate for the pages of his 'Selborne.' For these verses the curious reader may consult Mr. Bell's edition, vol. i. p. lii., and may add to the version there given the following concluding lines, which appear in a copy made early in 1789 by Thomas White—

" For Cloacina with resistless sway
    Demands her right, and authors must obey."

In July brother Thomas and Mrs. Henry White
and two daughters came to stay at Selborne.

To Mr. Churton.   Selborne, Aug. 4, 1788.

Dear Sir,—You must not expect that I who am at best
but a venerable vegetable, remaining like a cabbage on the
same spot for months together, should be able to furnish out
a letter full of entertaining incidents like you, who are
flying from diocese to diocese, and from cathedral to
cathedral. I thank you for your information respecting
Bourn Well-head, and might have made some use of it, had
not my last corrections, and *errata* been sent to London the
very day that I received your letter. The fate of my work
is now determined; and as the tree is fallen, it must lie.
My brother and nephew have spared no expense about it,
and particularly on the engravings, which have cost a considerable sum. This book will as you suppose not be
published now till the autumn, when the town begins to
fill. In the interim the author will be in no small a squeeze;
and will feel like a school boy who has done some mischief,
and does not know whether he is to be flogged for it or not.
As you were accessory to making me an author, you must
defend me if I am attacked unreasonably. 'Orna me,'
I think Tully says somewhere. As to your own work you
will, I find, spare no pains about it; so that not long hence
I expect to see you a very distinguished Biographer. As to
my beds they have all been full some time; and will be
more than full next week: however, by the end of the
month or the beginning of next there will be room for
you: so finish your N. excursions, and then come and stay
a good while. The heats are so severe that the reapers give
out and fall sick. As to Dr Chandler and Lady, they will

I think be quite roasted in their return from Naples.
Camden, or rather the Bishop [Gibson], says that there is
a medicinal spring at Bourn. I cannot but wonder at you
for losing your way ; a traveller so intelligent should always
be sure of his bearing. You talk of Archbishop Parker's
MSS. and the Coll. at Cambridge where they are preserved
with so much care ; but never mention what College
[Corpus Christi then called Ben'ct]. We have a fine
crop of wheat this year at Selborne and a good shew for
hops. The season becomes more and more sultry, and this
day my thermometer in the shade stands above 80°! Mrs.
Etty writes word that she is much pleased with the manners
and mode of life of the gentry of Cornwall. Her son
* . . . . . . Magd. Coll. He will have better luck,
. . . . . . Pray present my most respectful compliments to Dr Loveday, and lady unknown, and to
Dr Townson whom I once saw about 40 years ago. A
Mr. Headley of Trin. Coll. has published a collection of
ancient poetry in two volumes, and seems to intimate that
he may another day undertake more of the same kind. As
I can direct him to a book which probably he has never
seen ; and as that book was written in the time of Queen
Elizabeth and may contain words and expressions explana-
tory of passages in Shakespeare, I have thought of writing
to his friend, Mr. Benwell of Caversham, and of recommend-
ing that old poetry to their consideration. I remain,

Your obliged and

Humble servant,

GIL. WHITE.

My brother and I ride out every day when the heat will
permit. Family respects attend you. Thermometer 82°.

In August the house at Selborne was full of
relations, since Henry White came from Fyfield

---

* Letter imperfect.

with several of his family. He records in his *Journal* a fishing expedition to "Frensham great pond," besides several observations on natural history; and regretfully notes—

"Aug. 14. Selborne Hanger dreadfully denuded, and yᵉ grand Hanging Wood facing Bro: White's garden horribly mangled.

"[Aug.] 19. Famous Druidical Stones brᵗ fr. Bridestone Lane by Nepʷ Edmᵈ."

The era of improvements at Selborne was by no means over, since the *Naturalist's Journal* records—

"1788, Oct 10. Nailed up a Greek, and an Italian inscription on the front of the alcove [? the *new* Hermitage] on the Hanger.

"[Oct.] 27. Set up again my stone dial, blown down many years ago, on a thick Portland-slab in the angle of the terrass. The column is very old, and came from Sarson house, near Amport, and was hewn from the quarries of Chilmarke. The dial was regulated by my meridian line.

"Nov. 3. Bro[ther] Tho[mas] sowed many acorns, and some seeds of àn ash in a plot dug in Baker's hill.

"[Nov.] 26. Finished shovelling the Zigzag and Bostal."

On Dec. 3rd, 1788, Henry White wrote in his *Journal* at Fyfield—

"Hamper from London containing yᵉ Natural History of Selborne, presented by yᵉ Author. A very elegant 4ᵗᵒ with splendid Engravings & curious investˢ [investigations]."

To the Rev. R. Churton.　　　　Selborne, Dec. 3, 1788.

Dear Sir,—There is an old maxim, which poor dear Mrs. Etty now and then made use of, that when once "Stir up we beseech thee, O Lord, the wills of thy faithful people,"

&c., had been read and passed over, the festival of Xtmass
came creeping upon us before we could be aware. Being
reminded by this wise saw, I began to think that I would
write to neighbour Churton, and invite him to Selborne,
when your agreeable letter came in.

It is a very flattering account that you give of the
reception which my book met with at Caversham and your
lodgings. There is reason to wish that the work may find
many more such candid readers : if not, what is to become
of the Editors, who have spared no expense in *getting* it
up, and who have printed off a large impression ?

I am now reading every day your friend Dr Townson's *
discourses, which give me, as you engaged that they would,
singular satisfaction : there is an acumen, and nicety of
critical discernment, not often to be met with. In his
sermon, p. 282, I am particularly charmed with the author's
remarks upon the use Xt. made of his parables, and the
reasons why they were so nicely adapted to the taste of his
hearers !

We have just heard that Miss and Reb. Chase were on
the wing for India. Their motive must be, no doubt, a view
of settling in the married state. Celibacy has something
in it so abhorrent to the sex, that they will flie from pole to
pole to avoid it. However, let their fate be what it may,
I wish them happy.

Pray bring what you transcribe respecting the κορωνη and
χελιδων; some use may possibly be made of it. I rejoice
to hear that Dr Chandler is well. I most readily condole
with you on the sad calamity that has befallen at Windsor;
and pray to God that He will be pleased speedily to
restore the King to a right use of his faculties. Should
the nation be long deprived of one of its states so necessary
to the constitution, such a spirit of party, it is to be feared,

* 'Discourses on the Gospels,' by Thomas Townson, D.D., Archdeacon
of Richmond and Rector of Malpas, Cheshire.

THE GARDEN AND PARK OF "THE WAKES," SELBORNE, LOOKING TOWARDS THE HANGER

[To face p. 188, Vol. II.

will break forth, as may make what we remember of political struggles a mere *civil game* to what may ensue.

Mr. Loveday has just written me a letter, in which he says, " If in the perusal any things should occur worthy of remark, such observations shall be transmitted to Selborne." Now pray tell that gentleman that any strictures from such a quarter will be most gratefully received; and be sure to add, that could such have been obtained before publication, they would have been deemed inestimable. Pray come on the 24th; for if you cannot be as regular in your migrations as a ring-ousel or a swallow, where is the use of all your *knowledge*? since it may be outdone by *instinct*. When Lord Botecourt was Governor of Virginia, a slave, meeting him, pulled off his cap, and made him a bow, which the benevolent peer returned. Good God! says a by-stander, does your Lordship pay any regard to such a wretch? By all means, says the good nobleman: would you have me outdone in common civility by a negroe? Mrs. J. White joins in respects to you and J. Etty; and to Mr. Ventris, when you see him.

Y^r most humble servant,
GIL. WHITE.

On December 15th, 1788, Mulso, who had also received an early copy of his friend's book, writes :—

Dear Gil.,—I have longed to write to you some time to tell you how very handsome I think you in your new dress. But, alas, I and all my family have been ill with what is called Influenza: it seizes the weak part of every-body, and therefore varies with the constitution. . . . The time of unwellness is most agreeably filled up with a real good work, and especially when that book is the production of a well-beloved friend. I was much obliged to your brother Benjn for sending me your 'Selbourne' so early,

before it made its appearance in the world; I wrote to
your brother about it: he was very careful and kind, and
I wrote to my own brother to call on him and pay for
it, which he has promised me to do. And now as to
what I think of it; you have known long, as I have read
all the natural observations before, and given them such
commendations as I could give; which wanted such weight
as a thorough knowledge in that branch would have given,
and which the book deserves. As to the Antiquities, you
have given to them such a grace in your manner of treating
the subject, as would give a pleasure and a hunger of
reading to a man not an antiquarian. Your book was mentioned with respect by our Chapter (a full one), and the
volume ordered to be bought for the library. The Prints
do not satisfy me, nor do they do justice to your beautiful
scenery.

I do not know whether you will resent any fault being
found with the care of the printing; I have hardly ever
seen a book so well attended to, and so happily finished off.

An account which followed in this letter of the
pairing of certain jackdaws which frequented the
Deanery roof, opposite to the prebendary's study
window, may be pronounced, in this case, a very
strong, as well as very early, instance of that instruction in the observation of Nature, which the
book in question was to exercise in the future.

CHAPTER IX.

At last, then, the book, which was literally the work of a lifetime, was published.

Its reception was not equivocal. One of the most striking as well as prophetic testimonies to its value came, soon after its appearance, from Dr. Scrope Beardmore, Warden of Merton College, Oxford; who said to a nephew of the author's, "Your uncle has sent into the world a publication with nothing to call attention to it but an advertisement or two in the newspapers; but, depend upon it, the time will come when very few who buy books will be without it"—a prediction which has been singularly fulfilled.

Thomas White fulfilled his promise to review his brother's book in the January and February numbers of the 'Gentleman's Magazine' for 1789, pp. 60, etc., 144, etc.; his notice being written in very good taste, as coming from a brother, and not too laudatory.

The pleasure which the congratulations of his friends gave the author must have been, however, sadly marred by an occurrence which was nearly

coincident with the issue of his book to the public;
since, on December 28th, 1788, he lost his youngest
brother, Henry, who died very suddenly at Fyfield,
aged 55, leaving a numerous family.

On January 5th, 1789, Mulso writes :—

"Though letters of condolence are very ineffectual, yet
cannot I bear to pass unnoticed the sad event of which you
informed me in your last. I am struck with it as one of
the wonderful decrees of Providence, that a person on whom
so many other creatures depended for provision, comforts,
and education, should be so hastily struck out of the book of
life.

"Yours is the most happy family that I know in being able
to give mutual help on these necessary calls. I heard the
other day with great pleasure that Mrs. J. White's son is
much admired in his way, and gets into great business.

Mr. Lowther and Dr. Sturges (both able men) admire your
book, particularly the Natural History, which not only
seems well founded, but has an originality in the management of it that is very pleasing. I see that you avoid
naming names, yet when you are mentioning Sunbury,
and a friend that you visited there, I a little repine that
my name did not stand in a book of so much credit and
respectability; and I am ready to say with Tully, 'Orna
me.' "

To Thomas Barker.   January 8, 1789.

Dear Sir,—You must have heard no doubt before now
of the sad and afflicting news from Fyfield; of the sudden
and unexpected event that has plunged a numerous family
in the deepest sorrow and trouble. How the poor man
has left his concerns, and how the widow and children
are to proceed, I have not yet heard: however as money

will probably he wanted, my two brothers, and nephew
Ben., and self have each began with a present. When
the news arrived here, I wrote away immediately to Lady
Young,* entreating her to apply to the Chancellor for the
living of Fyfield for Sam[son], but she returned for answer
that she had kept up no acquaintance with Lord Thurlow,
though a near relation, for many years. Brother Ben.
wrote immediately to the Chancellor, and brother Thomas
applyed by means of Dr Lort, who prevailed on the Bishop
of Bangor, Dr Warren, to press the matter home, and to
move the compassion of the great man by representing
the afflicted situation of the family. For a short time
we were almost ready to flatter ourselves with some hopes
of success; the great bar seemed to be, that Samson was
not in orders : however two or three days ago a note came
to brother B. from the Chancellor informing him that
Fyfield was disposed of, but that Uphaven was at his
service. Now you must remember that Uphaven was a
very small vicarage indeed: however brother H., I hear,
had improved it not a little.

I now see more and more reason to be thankful to providence for enabling me to procure so many friends to assist
me in getting Sam. elected fellow. That young man, whom
all speak well of, may become the stay and support of the
family. By the statutes of his college he will not be able,
I fear, to take orders till June, when he may take possession
also of a fine curacy, now held for him. Charles† also is
intended for orders, and has kept some terms at Oxford.
It is needless to inform you that we experience a long
and severe frost, which commenced Novr. 23rd, and has
never been out of the ground since. The snow in this
district has been very little. After a very dry spring and

* Formerly Miss Battie.

† Henry White's second son, Charles Henry. He became Rector of
Shalden, Hants.

summer and autumn, about ten days in September excepted, the failure of water is remarkable. The ponds are all dry, and most of the wells in the village, and among the rest my own. As to Edmund White's tank, it has failed for these seven weeks; and he is obliged to fetch his water from the south side of Nore hill. My column of rain on the other side makes a very small figure, in respect to what I used to send you. Mrs. J. White returns my sister thanks for her late letter. I am disposing of her guinea among the poor. Never were gratuities of that sort more acceptable than now.

We hear now from all quarters that nephew John the Surgeon gets business very fast, and is allowed to be in a way to become the first medical man in Salisbury. Little Ben. thrives and grows very fast.

Mrs. J. White joins in best respects and wishes.

I am Y^r affectionate brother,

Gil. White.

In 'The Topographer for the year 1789,' etc. vol. i. part i., April, appeared a review of Gilbert White's book, from which the following extract is taken :—

"It is a part of my plan to notice all new publications, that illustrate the topography of the kingdom. And the first book upon this subject that it is our province to review becomes a very pleasing task to us, for a more delightful, or more original work than Mr. White's History of Selborne has seldom been published. . . . Natural History has evidently been the author's principal study, and, of that, ornithology is evidently his favourite. The book is not a compilation from former publications, but the result of many years' attentive observations to nature itself, which are told not only with the precision of a philosopher, but with that happy selection of circumstances, which mark the *poet*. Throughout therefore not only the understanding is informed, but the imagination is touched. And, if the criterion of excellence, that Dr. Johnson, I think, somewhere in his lives of the poets proposes be true (as it certainly is), Mr. White's book is excellent, for I beheld the end of it with the pensive regret with which a traveller looks upon the setting sun."

To justify his opinion the author of this notice proceeds to give certain selections from both parts of the book, noting that—" On p. 68 is a very elegant little poem, entitled ' The Naturalist's summer evening's walk,' which confirms what we have before advanced of the author's poetical powers."

To Mr. Churton.   Seleburne, May 20, 1789.

Dear Sir,—By my late letter from Dr Chandler still at Rolle, I am informed, that he and lady wish now to come home; but are prevented by their boy, who being with a wet nurse, cannot be moved, till he is weaned. The D^r desires to know the state of our neighbourhood; and wishes me to look for an house in the vale of Alton. I can find none in the environs, but in the town itself I have seen one, roomy, cheap, and fit for a gentleman's family. So I have sent an account of the matter to Rolle.

From the circumstances mentioned in your last respecting Dr Loveday's ponds, and from what I have heard from Rutland, and from my own observations at home I am convinced that the fish which were saved in some pools, were preserved by the rills and springs which ran through them. A current of water must constantly introduce a current of air: while in waters where there were no springs, the air is breathed over and over again till it becomes unfit for respiration. At the instance of Sir Joseph Banks, my brother-in-law, Mr. Barker, of Rutland, has just written a circumstantial account of the long frost last winter, which is to be sent to the members of the Academy of Sciences at Paris, who, it seems, have made similar applications to the *literati* in other parts of Europe. Mr. B. is of opinion that the usual method of breaking holes in the ice is pernicious.

You did well to be zealous for the friend of your friend : but were you apprized that there was to be no election at Oriel last Easter ? Mrs. Etty wrote at the same time on the same occasion.

Littleton Etty is appointed one of the Clerks of the General Post Office at the stipend of £40 per annum and Mrs. E. like a good mother, is going to live in London to superintend the morals of her son. If you had stayed three days longer at Xmas we had made a staunch ornithologist of you for life; for a young farmer brought us in two rare water-birds, a *Gooseander* and *Dun-diver*, a drake and a duck of the genus *Mergus merganser* Linn. The beauty of the colours, and the curious formation of the parts would have struck you with wonder; and would have obliged you to reflect on the wisdom of God shown in the works of Creation; and perhaps in no instance more, than in that of the birds of the air !

Dr Chandler has sent us several curious remarks made by him and his friends respecting the swallow kind even so far S. as Rome and Naples.

The world has been so indulgent to my book, that I begin to hope that the editors will be paid for the trouble and expense they have bestowed on the publication.

Mr. Gough's *Camden*,* I see, comes out in three vol. fol.

---

* Richard Gough (1731–1809), an English antiquary, began to prepare an edition in English of Camden's ' Britannia ' in 1773. The work appeared in 1789.

towards the end of this month: whether he would have taken any notice of the antiquities of Seleburne, had he seen them in time, before his account of Hants was printed off, I cannot pretend to say. He made honourable mention, I remember, of my account of the British *Hirundines* in his ' British Topography.'

Charles Henry, the second son of poor brother Henry White was ordained Sunday seven night by the Bishop of Wintoñ: as to Sam. he remains a layman still, on account of an awkward statute which forbids a fellow of Oriel from going into orders till he is regent Master.

While I was writing this the Reading *Mercury* brought in the news of the death of poor good Mr. Loveday of Caversham. He was gathered-in like a ripe stock of corn, in a good old age; having lived a blameless life; and been the occasion of many good and benevolent actions. You and a friend were I hear at " Horace's head " on the day of procession.* Nephew Ben. and wife have just left us. Brother Thomas and Mrs. J. White join in respects.

The bloom of apple is again prodigious !

Your most humble servant,
GIL. WHITE.

The *Naturalist's Journal*, under date May 24th, 1789, South Lambeth, contains a copy of a letter from Dr. Chandler, dated Rolle en Suisse, April 4th, 1789, regarding his observations of swallows, etc., there.

The publication of Gilbert White's book naturally brought him into relations with other students of natural history, hitherto strangers to him. One of the first of these to write to him was George Montagu (1755 – 1815), subsequently the well-

---

* Of King George III. to St. Paul's, to return thanks for his recovery.

known author of the 'Ornithological Dictionary,' and several other highly-esteemed works.

From Col. G. Montagu.

Easton Grey, nʳ Tedbury Glostershire,

May 21st, 1789.

Sir,—Although I have not the pleasure of being personally acquainted with you, yet I flatter myself you will pardon this intrusion of an enthusiastic ornithologist.

I have been greatly entertained by your 'Natural History of Selborn,' in the ornithological part of which, I find mention made of three distinct species of willow wrens. Can you inform me if they are (besides the common) the larger and lesser pettychaps of Latham, neither of which are described by Pennant in his 'Brit. Zool.'? he describes a species with the inside of the mouth red, which I can not make out in this country: those two of Latham I believe I have got, as far as I can judge from the description that gentleman favoured me with: but his sedge wren I am at a loss for, as he describes the sedge bird besides of the 'Brit. Zool.'

I should esteem it a particular favour, if you have it in your power, if you will favour me with the weight and description of the two uncommon willow wrens.

I was induced to take this liberty as you say you are a field-naturalist and perhaps may have it in your power to assist me in my present pursuit. I am collecting and preserving the birds and their eggs of these parts, a provincial undertaking in which I am got forward, and as those of Hampshire and Wiltshire are nearly congenial (the coast excepted) some species, I presume, are more frequently met with about you than with us: will you excuse my mentioning a few that should they fall in your way, you will confer a considerable obligation on me by favouring me with them.

The hawks and owls are difficult to get: of the former I want all except the sparrow, kestrel, and common buzzard; of the latter all the eared and the little owl. The great butcher-bird, and woodchat, goatsuckers, crossbill, aberdevine, or siskin, and spotted gallinule, with many cloven, and web-footed water birds, together with any of their eggs. And as you mention snipes and teals having bred near you, their eggs would be highly acceptable, with others not common which you may be able to obtain. And in return, sir, if there is any thing in my former, or future researches that can afford you any satisfaction, I shall with the greatest pleasure communicate.

That amiable and excellent naturalist Mr. Pennant has done me the honor to say I have discovered some things to him he was not before acquainted with; and I flatter myself I have other notes in store when I have more time to write to him more largely upon the subject; this you know is the busy season for a naturalist, and the days are not half long enough for me.

A fine morning called me from this and in my walk my ears discovered a note I had never before heard. I pursued it into the thick of a wood, and after much difficulty killed the bird as it was delivering its song (if it may be so called) from the branch of an oak tree: it proved to be a willow wren; its note was certainly very different from any I ever heard before, somewhat resembling a note of the blue titmouse, it was continued without variety, like the grasshopper lark but not quite so quick or shrill, nor of so long duration; between each song the pause was considerable: the note I confess has staggered me, but its appearance, size, &c. discover nothing new. The common willow wren, I well know, has two very distinct songs: the first after their arrival, before they are paired, I considered as their love-call, the other their soft courting or amourous song: as to shades of colour, or size, this species vary considerably:

that of the male is much stronger and brighter than the female, and is considerably larger, and even in the same sex there is frequently a visible difference. I last year killed a male and female together, when the former was in pursuit of the latter on her first arrival in the spring (as I suppose you know all our male migrative birds precede the females in their vernal visit) in these the great disparity of weight and difference in colour would have puzzled me exceedingly, had I not some time before the barbarous act was committed, paid attention to the addresses of the male. I confess I am not acquainted with the one you describe with the primaries and secondaries tipt with white, and if you are still of opinion that is a distinct species I should be obliged to you for it. If you should favour me with any small bird at this season, it will be advisable to wrap it up in soft paper sprinkled, or damped, with vinegar, first laying the feathers smooth and then cover'd with thicker paper, wetted with the same. This will both preserve the bird moist and defy putrifaction: larger birds should be carefully opened with a sharp knife from the vent upwards, laying the feathers back with damp paper to prevent their being blooded in taking out the intestines; a little alum or nitre should then be thrown in and the incision stopped with tow; and if a little of the alum and tow was put into the mouth it would ensure its coming to me in good order. If you have any conveyance to Bath and you will take the trouble of directing either a box or basket for me to the care of the R^t Honble Lady Jane Courtenay, Milsom Street, Bath, I shall get it the day after it arrives there.

Notwithstanding my post town is in Glocestershire, I live in Wiltshire, where I shall be happy to obey any commands from you and remain Sir

Y^r most obed^t humble serv^t,

G. MONTAGU.

A little later he wrote again to Selborne :—

From Col. G. Montagu.

Easton Grey, June 27th, 1789.

Dear Sir,—I am exceedingly obliged to you for your polite favor, and hope you will excuse the mistake in my address.*

I am not able to boast of being an ornithologist so long as you, though I have delighted in it from infancy, and was I not bound by conjugal attachment, should like to ride my hobby to distant parts, yet I agree with you, that naturalists in general attempt to explore too wide a field, their researches are too extensive; whereas if persons well qualified were to confine themselves to particular districts the natural history compiled from provincial authors would no doubt throw much light on the subject.

The difficulty of comparing birds by recollection, and the great similitude of some distinct species, made me attempt to preserve them; by which means I can at any time bring them together for a minute inspection.

I confess myself greatly obliged to your work for the discovery of the *third* species of *Willow Wren,* and for the first determined separation of the other two species, with whom I was perfectly well acquainted as to their notes, but suspected that the same bird might produce both notes promiscuously. Your work produced in me fresh ardor, and with that degree of enthusiasm necessary to such investigation, I pervaded the interior recesses of the thickest woods, and spread my researches to every place within my reach that seemed likely. I was soon convinced of two distinct species, not only in their song, but in size, colour, eggs, and materials with which they build their nests. The third species, which you seem to think is peculiar to your beech woods, I flatter myself I have at last discover'd to be an

* His first letter was addressed " Gil. White Esqre."

inhabitant of this part, but they are scarce and partial: three only have I discovered, two of which I brought down with my gun from the top of tall oak trees, in a thick grove interspersed with brambles. From their reiterated note, somewhat resembling the blue titmouse, and their colour being more vivid than the other species, I do not hesitate to pronounce it that discovered by you, though mine did not possess any white on the tips of the quill, or secondary feather, but the belly was of a pure white, and the action of its wings agrees with your description: besides the note it commonly uses which is somewhat grasshopper-like, it produces a shrill note five or six times repeated something like the marsh titmouse. One pair of these birds I only know of about this neighbourhood now, the nest of which I have not been fortunate enough to discover: if one should come across you it would be an acquisition to me. You are perfectly right in saying the name of *Willow Wren* is very inadequate. I wish you had given them distinct names, as I believe you have the merit of the original discovery. I am surprised Pennant makes no mention of these acquisitions to ornithology, as your letter of the 17th of Augt. 1768 long preceeded his last ed^n. Do you know if Latham has adopted them in his 'Systema Ornithologiæ,' which is to come before the publick next winter.

I am at a loss for your *blue pigeon-hawk* especially as you say its female is brown: from its place of resort I should conceive it to be the *Hen harrier* and that you had not corrected the mistake of other ornithologists, and which Pennant fell into in his first edition where he gave the *Ringtail* for its female. Their habits and manners are nearly the same only the latter perch on trees occasionally: its white rump at once distinguishes it from all others when skiming over the surface of the earth: like the Henharrier it makes its nest on the ground. Both these species we have, but not preserved, having not been able to procure them,

being scarce and shy; perhaps I may be favoured with them
from you, as well as their eggs, another season if not this:
If your Pigeon-hawk should be different I should be obliged
to you for further explanation as I am not acquainted with
it by that name.

The *Hobby*, which I want, has been called the blue hawk
by some; its eggs I should be glad of and are no doubt
to be found in your extensive woodlands: they are scarce
with us. You are surprised at my requesting of you the
Goat-sucker: 'tis true many parts of this county produce
them, but they are not to be commanded, and one bird
in the spring or before August is worth twenty after that
time, as most birds are then out of feather, and the young
ones are seldom in full, or proper plumage till the winter,
and many till the ensuing spring. In the latter end of
October birds have mostly done moulting, and are again
fit for preservation: however scarce birds are acceptable
at all times, till a better supplies its place. Since I wrote
I have killed the male Goatsucker, and as I have seen a
female it is probable I may get it, but the egg I despair
of in this part. You seem to suspect the distance through
town would endanger spoiling any specimen during warm
weather. 'Tis true without some little precaution it might,
but if a bird was carefully opened with a sharp knife from
the vent along the abdomen to the lower part of the breast
bone, and the bowels taken out (after laying the feathers
back with a little damp paper to prevent blooding them
in the operation), and a little powdered alum was put
therein and some tow, or soft paper to prevent the feathers
from falling in, and to soak up any superfluous moisture,
would ensure their passage twice as far; a little alum in
the mouth and throat might be added if the bird is stale
before sent: and if the weather is warm a brown paper
damped with vinegar would be an excellent second wrapper
—first laying the bird smooth in soft paper, which can

not be so well effected in stiff paper, then packed in a basket
or box with dry straw to prevent friction; there is no fear
of their coming safe to hand, even though the distance was
greater and more intricate.

From Alton to Town is one day's journey, from thence
to Bath one, the third day brings it to hand; and I am
apt to think consigning it to a friend in Town might rather
delay it than otherwise. If directed for me at Lady Jane
Courtenay's, Bath, through London, with perishable marked
on the direction or any other devise to hasten it, there is no
doubt of it coming to me on the third or fourth day.

I remain Dr Sir,

Your much obliged and faithful humble serv^t,

G. MONTAGU.

During the summer of 1789 many relatives visited
at Selborne on their way to and from Fyfield. On
August 16th, 1789, Mulso, who was detained by
ill-health at his Winchester house, writes:—

"How is your sweet retreat this year? What are your
enjoyments, and what friends have you about you? Let
me hear from you, my old friend, now and then; if I was
not solicitous about you, I should not deserve you. . . .
My sister Chapone goes to Bath on Thursday; she desired
me to ask you if you had read Dr Darwin's 'Loves of the
Plants.' She admires the poetry; but the subject, ah pah!
'with the loves of flowers,' says she, 'one might play with
one's fancy; but the loves of stamens and pistils is too
much for my strength.'"

A remark from which it may be inferred that "the
admirable Mrs. Chapone" was not a botanist.

"THE WAKES," SELBORNE, FROM THE VILLAGE STREET

[To face p. 204, Vol. II.

To the Rev. R. Churton.   Seleburne, Sept. 1, 1789.

Dear Sir,—Your letter of July 31st lies before me, and informs me that you are now breathing your native air, which, I hope, will agree with you: Malpas will moreover, I trust, prove a mother to you, and not a step-mother. The reason that Edmund White delayed his journey to Oxford was the badness of the weather, which broke-up the party; however, he went himself on the last day of term but one, and took his degree on the last day. I rejoice to hear that your good friend Dr Townson continues so well at his advanced time of life; and desire my respects to him. As to Dr Chandler I have heard from him twice in the course of this summer, and have looked him out an house, the best house in Alton: he seemed in his last to pay some attention to my information; but I have doubts about his settling, and do not depend on him as a neighbour. He at present is much embarrassed by the troubles in France, which would render a journey through that kingdom truly dangerous. He talked in his last of going up to Basil, and so down the Rhine to Holland. While I was in town I turned over Mr. Gough's 'Camden':* it is truly a Herculean labour: no wonder that there should be some mistakes. In the map of Hants I saw *Wetmer* Forest instead of *Wolmer*. Were I to live near you I verily believe I should make an ornithologist of you. I have just found out that the country people have a notion that the *Fern-owl*, or *Eve-jarr*, which they also call a *Puckeridge*, is very injurious to weanling calves by inflicting, as it strikes at them, the fatal distemper known to cow-leeches by the name of puckeridge. Thus does this harmless, ill-fated bird fall under a double imputation, which it by no means deserves,

* Camden's 'Britannia' had long been a familiar book to Gilbert White, since a copy, published in 1695, edited by Gibson, had been presented to him in his twentieth year by "the Rev. Mr. Brown, Vicar of Bray," as he records on a flyleaf.

in Italy, of sucking the teats of goats, where it is called
Caprimulgus; and with us of communicating a deadly
disorder to cattle. But the truth of the matter is, the
malady above mentioned is occasioned by the *Œstrus bovis*,
a dipterous insect, which lays its eggs along the backs of
kine, where the maggots, when hatched, eat their way
through the hide of the beast into the flesh, and grow to
a large size. I have just talked with a man who says he
has been called in, more than once, to strip calves that
had died of the puckeridge; that the ail or complaint lay
along the chine, where the flesh was full of purulent matter.
Once I myself saw a large rough maggot of this sort taken
out of the back of a cow. These maggots in Essex are
called *wornils*. The least attention would convince men
that these birds, weak and unarmed as they are, cannot
inflict any harm on kine, unless they possess the powers
of animal magnetism, and can affect them by fluttering
over them. Pray ask your brother whether he knows the
bird and the distemper, and whether Cheshire men are
perswaded that the latter is occasioned by the former. We
had experienced a most lovely wheat-harvest; but now
there is rain, which will respite the partridges for one
day at least. As soon as we came from town my house
became full of visitors; we have had Mr. and Mrs. Sam
Barker from Rutland, and Miss Eliz. Barker, a fine young
woman, who is allowed to be a very good lesson-player on
the harpsichord. They left us last Tuesday. We now
expect my brother Thomas White and family. My brother,
I hear, is very well. Pray present my respects to D^r
Loveday, and tell him I should be very glad to see any
notes or remarks made by him or his venerable father on
the history of Selborne: could they have been procured
before publication, they would have been more valuable,
because I might then have availed myself of their corrections. My book is still asked for in Fleet Street. A

gentleman came the other day, and said he understood that there was a Mr. White who had lately published two books, a good one and a bad one; the bad one was concerning Botany Bay,* the better respecting some parish. The bookseller recommended the parochial work; and told the enquirer that he did not believe the author ever had been at Botany Bay, or had ever written about it.

Mrs. J. White joins in respects. Mr. and Mrs. Edmund White are gone to Ramsgate in Kent, a watering-place on the coast. Mr.† and Mrs. Taylor are here. We have again a very fine crop of hops.

Yʳ most humble servant,

GIL. WHITE.

From the *Naturalist's Journal*—

"1789. Oct. 3. B. Th. White sowed two pounds of furze seed from Ireland on the naked part of the Hanger. The furze seed sown by him on the same space in May last is come-up well.

"[Oct.] 22. Bro. Tho. White sowed the naked part of the Hanger with great quantities of hips, haws, sloes and holly-berries."

The labour was, however, useless, since in December, 1790, it is recorded that the "sheep have browsed on them as fast as they sprouted."

The following letter contains a favourable mention of his friend's book :—

From the Rev. R. Churton.

Brasen-Nose, Oct. 25, 1789.

Dear Sir,—The date of your last and still unanswered letter I am ashamed to mention. However, though I have

---

* 'A Voyage to Botany Bay,' by John White, a book in much request now, and never a bad one. As a matter of fact it was not published at this time, and bears date 1790. Of course, it had been advertised before.—A. N.   † The Vicar of Selborne.

not written to you, I am glad to hear my friend Miss Reeve
has been seeing you.   Very learned and, I hope you think,
very civil, a knight's eldest daughter with perhaps a thousand
pound for every year of her age, or at least half as many.
Hendon House near Maidenhead is in a most charming
country, and as yet perhaps a *non-descript*.   As you are
perfectly acquainted with every quadruped and bird and
insect and flower near Selborne and have introduced them to
the public and to immortality, it will be a pleasant circumstance to vary the scene, and add celebrity to Windsor and its
neighbourhood.

> " Methinks I see thee straying on the " thicket,
> " And asking every" bird that roves the sky
> " If ever it have" seen fair Selborne's down.

I cannot say but I am interested in this expected
migration.   I can then whip over to see you often and
take a dinner or a bed for a single night and return to
college.   But Selborne is a long way off.   And yet it is
worth going a long way to see, if it agrees at all with
the account which a very curious and interesting book in
my room gives of it.   You must know that I am reading
this work with great avidity in the very few leisure moments
that I can find or steal, and I am only sorry that the
Index to a volume containing such a variety of useful
and authentic information is not much more copious.   If
you are acquainted with the writer of this " good book,"
you may tell him, with my humble service, that I hope
to be able to give him some papers that may help in the
second edition to remedy this single defect.   But it is
time to answer your queries in regard to the distemper
called " *Puckeridge*."   I consulted my brother and other
persons on this subject and minuted down the particulars
he gave me, in which others also concurred.   The name
of *Puckeridge* is unknown in Cheshire.   The disease along

the chine, or rather the maggots that cause it, they call "worrybrees" and a single one "worrybree." But they are so far from thinking these maggots prejudicial, that, on the contrary, they judge the calf that has these "worrybrees" in the back less likely to be *struck* (as they call it) with the *hyant*, which is or is considered a distinct disorder. When they are affected with this it is perceivable by the hand; for the skin is hard, and rustles (if you know that word) under the hand when rubbed by it. Sometimes there is one or more spots of this nature, and sometimes the body is almost covered with them. When the skin is taken off, the flesh in those parts is like jelly. It is deemed almost incurable, and they die in a few hours. My brother never knew or heard of more than one instance of a calf thus stricken recovering. That was but slightly affected, perhaps in a single spot; and the owner took the skin off the part and put in a rowel, or something of the sort. This disorder prevails most in Spring and Autumn, and commonly in calves of the first or second year, seldom in older cattle. Quid existimas de hac questione, an Puckerigium sit Hyantium? and whence comes this remarkable word? Are the Hyades supposed to cause it? I have heard the expression planet-struck, but whether of this disease I am not sure. In Cheshire they call calves the first winter *twinters*, in the second year *sterks*. The last is common, the other growing obsolete. I take it to be a contraction of *two winters*; for it is applied to them not as soon as calved, but when, if they were calved in winter, they are two winters old.

Dr Loveday had a letter, about six weeks ago, from Dr Chandler, still at Rolle, but talking of moving, but yet, if possible, more unsettled in his plans than ever. You mention jack-daws building in rabbit-burrows. It is not equally extraordinary, but perhaps you may not know that they build in Elden hole, a perpendicular aperture in a rock, about 90 yards deep, in Derbyshire. I did not take any of their nests, nor, indeed, did I see any; but I heard them chattering most loquaciously, and perhaps "disturbed their ancient solitary reign" by throwing stones into their little kingdom, when I was in Derbyshire about 5 years ago. I go to town on Saturday and return the Monday se'nnight. I shall probably hear of you in Fleet Street, and in a short time, I hope (though I am unreasonable to expect it), be favoured with a letter. You will be so good as to remember me with my best wishes and respects to Mr. T. White, who, I understand, is now with you, as also to Mr. Edm. White, &c.

I am, dear Sir,

Your sincere and much obliged humble servant,

R. CHURTON.

Dr Bostock has gained a Chancery suit and another son. Remember me to Miss Reeve when she calls next.

To the Rev. R. Churton.

Seleburne, Dec. 4, 1789.

Dear Sir,—Tho' Oxford appears to my timid apprehensions to recede every year farther and farther from Selborne; yet to you who are in the prime and vigour of life, Selborne ought not to be one inch more removed from Oxford than when I first knew you; therefore we shall depend much on seeing you at Xtmass as usual. I have much to say to you: for surely we live in a most eventful and portentous period; when wars, devastations, revolutions, and insurrections crowd so fast upon the back of one another that a thinking mind cannot but suppose that providence has some great work in hand! But of all these strange commotions, the sudden overthrow of the French despotic monarchy is the most wonderful—a fabrick which has been now erecting for near two centuries, and whose foundations were laid so deep, that one would have supposed it might have lasted for ages to come: yet it is gone, as it were, in a moment!!*

These troubles naturally put me in mind of Dr Chandler, who, the last time we heard of him, was in Brussels, in a most uncomfortable situation, having his baggage seized and his papers tumbled about, for which he was in great concern. A man of his resolution and address, and who, by his long voyage to the Levant, has, as it were, been inured to dangers and difficulties, might by himself make his way through all the misrule and uproar that prevail in all the provinces of the Netherlands: but the case is very different where a man has a wife and infant to protect and take care of; and therefore I heartily wish that he and his family were safe at home. My account of our visit from Miss Reeve, who paid us a great compliment, and did us much honour, I knew would make you and Mrs. Ventris smile. I could tell you also if I had a mind, of a great honour received from Lady Coterel Dormer. You are very kind in taking the trouble, amidst all your busy hours, of enlarging my index: when I had carried it to its present bulk, I desisted out of pure modesty, thinking I should swell the volume unreasonably; but to say the truth when I showed it to my brother he expressed a wish that it had been fuller: it was then too late.

Your *worry bree* is undoubtedly a corruption of *breeze* or *breese*, a synonymous term with the *gadfly*, well known to naturalists: as to *hyant*, we know nothing of the term, or of the distemper intended thereby. When I was at Elden-

---

* Of all criticisms of Gilbert White's book the remark that "he was more concerned with the course of events in a martin's nest than with the crash of empires" has always seemed very inapplicable. It is true that he does not appear to have taken any very absorbing interest in these great events, but one would not exactly expect to find them treated of in what was professedly a natural history of a parish. His *Naturalist's Journal* occasionally contains entries of current events, such as the surrender at Saratoga, the sailing or return of naval expeditions, the loss of the *Royal George*, etc.

hole I remember to have seen daws flying out from that horrible and tremendous chasm. These birds, thought I, are wise in their generation: for here they breed uninterrupted from age to age, since the most roguish boys dare not interrupt their ancient inaccessible kingdom.

Are you a Whiteist, or a Badcockist? * for I hear every man in Oxford must be one or the other. I can tell you how you may do Edmund White a good office. When he and his wife were in Oxford, last summer, they quartered at the Bear-inn, where they left behind them the first volume of Dilly's prose Elegant Extracts. It is a very oddshaped volume in quarto, somewhat like a music book. If you could recover this book, it would be received with thanks.

Mrs. J. White and I join in respects to you and James Etty; and in best wishes to Mr. Ventris, who, we hope, is recovering his health and strength very fast. When does Bishop W. Smith, your founder, appear? We long to see you a biographer and to read the result of your painful and curious enquiries.

Y^r obliged and humble servant,

GIL. WHITE.

When you write, present my respects to Dr Loveday and Dr Townson. How I wish that we had such a man as either of them living at Selborne!

From the Rev. R. Churton.

Brasen Nose, Dec. 13, 1789.

Dear Sir,—Your excellent letter deserves a much better answer than I have time or ability to honour it with. But

---

* The Rev. Joseph White, Regius Professor of Hebrew and Laudian Professor of Arabic, was appointed Bampton Lecturer in 1784, whereupon he preached, before the University of Oxford, a set of sermons, comparing Christianity with Mahomedanism, which were printed in 1784, and soon reached a second edition. Upon the death, in 1788, of Mr. Badcock, a learned dissenting minister, it was discovered that a considerable share of the sermons was of his writing.

I can assure you of one thing, which you, in your kindness
to your friends, will be glad to hear of. I depended upon
having the pleasure, V.D., of spending my Christmas at
Selborne before your obliging invitation arrived, and on that
account declined Dr Loveday's invitation to pass the holidays
at Williamscot, where, however, I hope to be for two nights
towards the latter part of this week, and then, after speaking
twenty pounds worth of Latin on St. Thomas's day, and
eating mince pies with the Principal, to set off for Reading,
Tuesday the 22nd, and proceed for Selborne next day. So
far so good. But this is not all. I inquired for the volume
left at the Bear; and it is no discredit to the house that the
book was found safe in a drawer in the bar, and is now safe
in my room waiting to be put up in my portmanteau.
Dr Chandler Wife and son arrived at Clapham about a
week ago safe and well, as you will probably have heard
by some means before this reaches you. Alas! I have
only found time to read, and with much satisfaction,
the 'History of Selborne,' but not to do much in enlarging
the Index. However, the loss is less material as Dr Loveday
has already or will soon undertake it, and do it effectually.
Marvellous indeed is the state of things on the Continent,
and when and how good order and good government will be
restored is far beyond my ken. But an all-wise Providence,
which can controul the madness of the people, superintends
the whole, and seems, as you justly remark, to have some
great work in hand. I did not know till you told me that
the "fatherlanders," as the papers call them, seized D^r
Chandler's portmanteaus; and I was afraid they were lost
through negligence. I hope they were restored; but I have
not positively heard so. I shall be glad to learn the
particulars of the honour received from Lady Coterel
Dormer, and other matters, *ex ore tuo.* And among these
I am curious to hear more about *worry breese* and *hyant*;
for if the distemper known in Cheshire by the latter name

never visits Hampshire, the reason is well worth enquiring after. I scarcely know whether to call myself a "Whiteist" or "Badcockist." The pamphlet of Dr Gabriel I think clearly shews that considerable assistance was received, but by no means ascertains the degree. In my own notion the Professor would do well to state fairly and explicitly what was composed by Mr. Badcock, and what by himself; and there are also some circumstances in his behaviour respecting the note, which should be stated in a more favourable way to his character, if they can consistently with truth.

Bishop Smith sends his compliments, and thanks you for your kind enquiries; but he says he shall not "walk the town numbering good intellects" before next winter. His biographer has lately had so much unavoidable business on his hands respecting the living that he has no time to talk with the dead. I am, dear Sir,

Your very sincere and obliged humble servant,

R. Churton.

CHAPTER X.

To Samuel Barker.

Selborne, May 6, 1790.

Dear Sir,—We had heard that Mr. Haggitt * had been very ill; but were not aware, till your letter came, that his disorder was of so dangerous and alarming a nature. On his own account, and for the sake of his numerous family, we hope it will please God to restore him to his former health, and preserve his life for many years.

The Major Jardine that you mention was well known to my Bro. John, an active, lively, intelligent Scotchman, that had been a private in the artillery; but having had some education was ready to enter into any pursuit where knowledge was to be acquired. He showed a great facility in modern languages, had a taste for music, and a smattering in astronomy, &c., was good-natured, clever, ready to assist, communicative, and pleasant; but exceedingly poor, having married a Spanish girl without a farthing, who brought him a housefull of children, and all her hungry relations to live on him, when he was only lieutenant. He was supposed to be the son of a Knight of Malta, whom he called his uncle. What can be the meaning of the following advertisement, which I have seen in the papers? "The life of the Hon. Thomas Chambier Cecil, late knight of the shire for Rutland, father of the present member for

* Samuel Barker's father-in-law.

215

Stamford, and brother to the Earl of Exeter." What makes
me wonder is, because this man was always represented
formerly as little better than an idiot! Now you talk of
biography, have you seen the life of Mr. Elwes, late member
for Berkshire?

Dr Chandler and Lady, who have been abroad almost
four years, and who returned from the continent only last
February, have borrowed Selborne parsonage-house for the
summer, and came to reside last week. The D^r who is an
unsettled man, likes this method of procuring an habitation,
because it looks so like *not* settling. Roaming about becomes
a habit with Gentry, as well as mendicants; who, when they
have once taken up a strolling life, can never be perswaded
to stay at their own parishes. The Lady is very big with
child, and sent for her midwife this morning: so they
reached Selborne just in time. They brought a little son
with them, a pretty boy, who was born at Rolle in Switzerland, as it were by accident, while they were posting home
for England. The D^r seems to like his child better, because
he is not sure in what kingdom he was begotten, whether at
Naples, or at Rome, or at Florence, or where. Rome is the
place that the D^r admires, where he can have his fill of
Virtù: he has, I find, secret languishings to return to that
capital; to study in the Vatican, and to dine with Cardinals.
In his passage to Italy they hired a ship at Marseilles, which
was to land them at Civita Vecchia: for some time they had
such prosperous gales that the master told them they would
be at their destination presently. But as they approached
Italy such squalls came off from the Apennine, that after
beating about for some days, and fearing that they must
have run for some harbour in Sardinia, they with difficulty
made Porto Longone in the isle of Elba. Their return from
Rolle in November last was singular enough. Not daring to
venture through France, they set out for Basle: here they
went 50 miles to the right to see the falls of Schaffhausen!!

When the D^r came to enquire of the watermen at Basle
what small craft they had on the Rhine, and whether any
house-boat; they said there was nothing but some very
small flat-bottomed wherries: but that they could tack
two of these together. On two such wallnut shells tyed
together embarked the D^r and Lady, the nurse and child, and
the French valet, without oar or sail, or any awning that
could be kept up; and thus ran at the rate of near 80 miles
a day to Dusseldorf, amidst the damps and fogs of Novemr.
on the expanded face of the Rhine, which was very full and
very rapid!! Here they turned off for Brussels, not being
aware of what was to befall them; but soon found themselves in a city that expected every day to be cannonaded
with hot balls. Here they stayed till they saw the streets
barricaded and intersected with deep entrenchments; and at
last escaped to Lisle, which was not without its difficulties
and embarrassments. The D^r and Lady went twice by water
down the Rhosne from Lyons: the scenery on the banks is
grand and beautiful. I have just received a letter from the
Revd James Anderson, LL.D., F.R.S., F.A.S. of the academy of
Arts &c. of Dijon &c., he directs from Edinburg, and having
seen my book desires my assistance towards his Bee, a
weekly work which he proposes to send forth as soon as
he can settle a correspondence to his mind. His prospectus
to his work is curious, and promises information. Nephew
John White of Sarum has got him an house, and two pupils.
Nothing but want of health will hinder that young man
from being successful, and prosperous. His business encreases. Mrs. J. White joins in best respects to yourself
and Mrs. Barker. We expect brother Thomas next week.

Your affectionate Uncle,

GIL. WHITE.

Mrs. Chandler is a pleasant woman with a good person:
while I was writing she was brought to bed of a daughter.

Respects at Lyndon.

On July 4th, 1790, a note in the *Naturalist's Journal* draws attention to "p. 274 [original edition, Letter LVI. to Barrington] of my natural History," in which he had noted that instinct varied and conformed to the circumstances of place and period. The entry continues—

"In confirmation of what has been advanced above, there are now two martins' nests at Tim. Turner's, which are tunnel shaped, and nine or ten inches long, in conformity to the ledge of the wall of the eaves under which they are built."

The next letter introduces a new correspondent in the well-known Robert Marsham (1708–1797), a gentleman much interested in arboriculture, who was living on his paternal estate of Stratton-Strawless, near Norwich. He had communicated papers on the growth of trees to the 'Philosophical Transactions' of the Royal Society, of which he was admitted a Fellow in 1781. Altogether, Marsham wrote ten letters to Gilbert White, of which the first is here given. Though the others are in the present writer's possession, it hardly seems necessary to print them here, especially as Mr. Bell has done so. Of the replies from Selborne ten have been preserved, and are now printed. It will be seen that the correspondence was only terminated by the death of Gilbert White.

From R. Marsham.   Stratton, near Norwich,
                                                July 24, 1790.

Sir,—I have received so much pleasure and information from your ingenious Nat. Hist. of Selborne, that I cannot

deny myself the honest satisfaction of offering you my
thanks: and I hope you will excuse the liberty that I have
taken. I have kept a poor imperfect journal above 50 years;
but it has been chiefly confined to the leafing and growth
of trees; and was undertaken by the advice of my most
estimable friend the late D^r. Hales. By that I find that
Linnæus's disciples, and their followers, are mistaken in
their supposed rule of Nature, *that all plants must follow in
order*. For you see by the Indications of Spring in the last
Vol. of the Phil. Trans. which very imperfect as it is, the
R. S. did me the honour to print, there are reverses of many
days.

Sir, I was much pleased with your Poetry in the Summer
Evening walk.—I hope you will excuse my asking you some
questions for my information. The copulation of Frogs as
you describe, is the manner of Toads with us: and I never
saw Frogs so engaged.

By your account of the Swallows on the 29th of Sep.,
1768, I presume that you believe in their migrating: and
there are very strong reasons to believe so of some other
Birds. Many Woodcocks are found by the Light-houses in
Norfolke in the Autumn, that are kill'd by flying against
the Lights: and the Earl of Orford informed me that the
Landgrave of Hesse sent him a ring taken from the leg of
an Heron with Ld. O. name upon it. This is certain proof
of the Heron's going from England: and myself have seen
(coming from Holland) a Wagtail (Motacilla alba) flying
about the Ship, seemingly at ease, when out of sight of
Land. These, without Admiral Wager's, Adanson's and
Smith's, (the earliest account that I recollect in print) are
sufficient for migration: and the proofs for torpidity are
also undoubted. So we may conclude they are both true.
But the annual increase in the Swallow tribe, which are lost
in Winter, affords unaccountable difficulties to be cleared.
I have had 4 pair attending my house as many years as

I can remember. If these produce two broods of 5 young, you see, Sir, one pair only, will in 7 years produce above half a million, 559,870 birds: yet the number every Spring appears the same. If both broods are destroyed, surely the old birds would be lessened by accidents, so as to be perceptible. If the early, or the latter brood is preserved, you see the next Spring Birds will be as 5 to 2, if all the old Birds are lost: and I never heard that Swallows are increased in any part of the Globe. We know that all the carnivorous Birds drive off their young as soon as they are able to provide for themselves; and I conclude that fish-eating Birds do the same: for when I was on the charming Lake of Killarney, I was told that was the case of a pair of Ospreys, that yearly nested on an Island of Rock in that Lake. But we cannot suppose the Swallow tribe can fear the want of provision. Sir, you know the Fern Owl is one of the Spring Birds, and appears here as the latest comer. I used to have many in my Woods; but since the long and severe Winter of '88 I have had very few. Is not this a presumptive proof of their torpidity? and that they were destroyed by the severity of that Season?—Your account of the 26th and 27th of March in 1777 was felt here in Lat. 52.45° but no swallows appeared. The 27th was insufferably hot, with a S.W. Wind; which changed in the afternoon to N.E. with a thick Sea-hase, and my Thermometer sunk above 20 degrees in 3 or 4 hours. The greatest change I have ever observed.—I find in 1776, Jan. 31st, your Thermometer sunk to 0°, mine of Farenheit was at 16°, and in 1784, Decr. 10th, when your Dollands was 1° below 0°, mine was but at 10°. The coldest Air I have measured was Jan. 19, in 1767, when it was down to 1°. I take the liberty to tell you this, as it possibly may be entertaining to you to see the difference of less than 2 Degrees of Lat.

Sir, when you print a 2ᵈ Edition, (which the merit of your Book will certainly soon demand) I hope in your description

of the Holt Forest, you will pay a compliment justly due, to
the Oak by Ld Stawel's Lodge: as I suppose it the largest
in this Island. I went from London on purpose to see it in
1759, and again occasionally in 1778. 'Tis at 7 feet full
34 feet in circumference, and had not gained half an inch in
19 years, yet I could not see it was hollow. If I measure
right, I make 14 feet length of the Holt Oak, to contain
above 1,000 feet, viz. above 320 feet more than the Cowthorp
Oak, which Dr Hunter in his Edition of Evelyn's 'Silva,'
calls the largest in England. I early began planting, and an
Oake which I planted in 1720, is at one foot from the earth
12 feet 6 inches 0 round; and at 14 feet (the half of the
timber length) is 8.2.0. So measuring the bark as timber,
gives 116F. ½ buyers measure. Perhaps you never heard
of a larger Oak and the planter living. I flatter myself,
that I increased the growth by washing the stem, and
digging a circle as far as I supposed the roots to extend,
and spreading saw-dust &c., as related in the Phil. Trans.
—I wish I had begun planting with Beeches (my favourite
Trees as well as your's) and I might have seen large trees of
my own raising. But I did not begin Beeches 'till 1741,
and then by seed; and my largest is now, at 5 feet, 6.3.0
round, and spreads a circle of + 20 yards diameter. But
this has been digged round and washed, &c.—The last
Winter was so very mild with us, that the leaves of many
of my very young Oaks preserved their green into April,
and a large Hawthorn (headed the preceding year) has its
old leaves now; which I never observed before, in any
deciduous trees: tho' I once had a second leafing of a
Hawthorn about Xmass. But those leaves faded before
Spring, I sent the account to Sir J. Pringle when P.R.S.
but he thought it not strange. Sir, if you do not take the
Ph. Trans. if you please I will send you a copy of my
'Indications of Spring,' as it may be an amusement to you,

to see how much later we are in Norfolk than you are in
Hampshire. I am, with great esteem,

Sir, your most obedient

humble servant,

R. Marsham.

P.S. I have now in a Stack of Blocks a young Cuckow
fed by a water-Wagtail.

To R. Marsham.

Selborne, near Alton, Hants,

Aug. 13th, 1790.

Good Sir,—As an author I have derived much satisfaction
from your kind and communicative letter, and am glad to
hear that my book has found its way into Norfolk, and that
it has fallen into the hands of so intelligent, and candid
a reader as yourself, whose good word may contribute to
make it better known in those parts. I am glad that you
happened to mention your most estimable friend, the late
Dr Steven Hales; because he was also my most valuable
friend, and in former days near neighbour during the
summer months. For tho' his usual abode was at Tedding-
ton; yet did he for many years reside for about two months
at his rectory of Faringdon, which is only two miles from
hence; and was well known to my Grandfather, and Father,
as well as to myself. If I might presume to say that what
you see respecting the copulation of toads is, I think, a
mistake, you will pardon my boldness: because the amours
carried on in pools and wet ditches in the spring time are
performed by *frogs*, which are more black and bloated at
that season than afterwards. As to toads they seem to be
more reserved in their intrigues.

With regard to the annual encrease of swallows, and that
those that return bear no manner of proportion to those
that depart; it is a subject so strange, that it will be best

for me to say little. I suppose that nature, ever provident, intends the vast encrease as a balance to some great devastations to which they may be liable either in their emigrations or winter retreats. Our swifts have been gone about a week!; but the other hirundines have sent forth their first broods in vast abundance; and are now busied in the rearing of a second family. Myself and visitors have often paid due attention to the oak in the Holt, which ought indeed to have been noticed in my book, and especially as it contains some account of that forest. You have been an early planter indeed! and may safely say, I should think, that no man living can boast of so large an oak of his own planting! As I had reason to suppose that actual measurement would give me the best Idea of your tree, I first took the girth of my biggest oak, a single tree, age not known, in the midst of my meadow: when though it carries a head that measures 24 yards three ways in diameter; yet is the circumference of the stem only 10 ft. 6 in. I then measured an oak, standing singly in a gentleman's outlet at about two miles distance, and found it exactly the dimensions of yours. After such success you may well say with Virgil,

"Et dubitant homines serere, atque impendere curas?"

In an humble way I have been an early planter myself. The time of planting, and growth of my trees are as follows: Oak in 1731—4 ft. 5 in. Ash in 1731—4 ft. 6½ in. Spruce fir in 1751—5 ft. 0 in. Beech in 1751—4 ft. 0 in. Elm in 1750—5 ft. 3 in. Lime in 1756—5 ft. 5 in.*

Beeches with us, the most lovely of all forest trees, thrive wonderfully on steep, sloping grounds, whether they be chalk or free stone. I am in possession myself of a beechen steep grove on the free stone,† that I am persuaded would

---

* The *Naturalist's Journal* of August 3rd records these measurements, giving also the site of the trees. "Oak by alcove, ash by ditto, great fir Baker's Hill, lime over at Mr. Hale's."

† Sparrow's hanger, at the south end of the village.

please your judicious eye; in which there is a tree that measures 50 feet without bough or fork, and 24 feet beyond the fork: there are many as tall. I speak from long observation when I assert, that beechen groves to a warm aspect grow one-third faster than those that face to the N. and N.E. and the bark is much more clean and smooth. About thirty or forty years ago the oaks in this neighbourhood were much admired, viz. in Hartley Wood, at Temple, and Blackmoor. At the last place, the owner, a very ancient Yeoman, through a blameable partiality, let his trees stand till they were *red-hearted* and *white-hearted* 3 or 4 feet up the stem. We have some old edible chest-nut trees in this neighbourhood; but they make vile timber, being always *shakey* and sometimes *cup-shakey*.

As you seem to know the *Fern-owl*, or *Churn-owl*, or *Eve-jar*, I shall send you, for your amusement, the following account of that curious, nocturnal, migratory bird. The country people here have a notion that the *Fern-owl*, which they also call *Puckeridge*, is very injurious to weanling calves by inflicting, as it strikes at them, the fatal distemper known to cow-leeches by the name of *puckeridge*. Thus does this harmless, ill-fated bird fall under a double imputation, which it by no means deserves—in Italy of sucking the teats of goats, where it is called *Caprimulgus*, and with us, of communicating a deadly disorder to cattle. But the truth of the matter is, the malady above-mentioned is occasioned by a dipterous insect called the *œstrus bovis*, which lays its eggs along the backs of kine, where the maggots, when hatched, eat their way through the hide of the beast into its flesh, and grow to a large size. I have just talked with a man who says he has been employed more than once in stripping calves that had dyed of the *puckeridge*; that the ail, or complaint, lay along the chine, where the flesh was much swelled, and filled with purulent matter. Once myself I saw a large, rough maggot of this sort squeezed out of the back of a cow. An intelligent friend informs me, that the disease along the chines of calves, or rather the maggots that cause them, are called by the graziers in Cheshire *worry brees*, and a single one *worry bree*. No doubt they mean a *breese*, or *breeze*, the name for the gad-fly, or *œstrus*, the parent of these maggots, which lays its eggs along the backs of kine.

But to return to the fern-owl. The least attention and observation would convince men that these poor birds neither injure the goat-heard, nor the grazier; but that they are perfectly harmless, and subsist alone on night-moths and beetles; and through the month of July mostly on the *scarabœus solstitialis*, the small *tree-beetle*, which in many districts flies and abounds at that season. Those that we have opened have always had their craws stuffed with large night moths and pieces of chafers: nor does it any wise appear how they can, weak and unarmed as they are, inflict any malady on kine, unless they possess the powers of animal magnetism, and can affect them by fluttering over them. Upon recollection it must have been at your house that the amiable Mr. Stillingfleet kept his *Calendar of Flora* in 1755. Similar pursuits make intimate and lasting friendship. As I do not take in the R. S. T. I will with pleasure accept of your present of a copy of your ' Indications of Spring.' Hoping that your benevolence will pardon the unreasonable length of this letter, on which I look back with some contrition, I remain, with true esteem,

Your most humble servant,

GIL. WHITE.

Any farther correspondence will be deemed an honour.

Marsham replied to the above letter on August 31st, 1790, and received an answer from Selborne dated October 12th, but this letter is unfortunately missing.

He wrote again to Gilbert White on December 29th, 1790.

On October 2nd, 1790, Lord Stawell sent the (now well known) hybrid bird, which was described in the *Naturalist's Journal*. It is here recorded that upon Gilbert White's recommendation "Mr. Elmer of Farnham, the famous game painter," was employed "to take an exact copy of this curious bird." The picture, which was given to the Naturalist by Lord Stawell, was engraved for 'A Naturalist's Calendar,' etc., published in 1795, and is now in the possession of a member of the White family. The bird itself found its way into the Earl of Egremont's museum at Petworth, where Mr. Herbert saw it in 1804. With the rest of this collection it has long ago perished.

On December 15th, 1790, Mulso wrote, in reply to a letter of inquiry from his friend, to announce his wife's death, "You knew her, my good friend, and you valued her as she did you." He continued—

I hear that, bating your deafness, you are in great soundness of body and mind. You have given in your work a very pleasing occupation for the last, in everybody. It is everywhere spoken of, and with the highest praises. Among others, Dr Warton is excessively pleased with it. Your nephew John called on me some time ago, and of him I enquired much after you. Alas, my good friend, how should we now do to converse if we met? for you cannot hear, and I cannot now speak out. I hear very good accounts of John White's success, and very satisfactory conduct in practice and behaviour, and that he has made a wise Partnership. Pray do you go this year to S. Lambeth? and at what time? How many branches have you to look after in every place that you go to! . . .

I am, my dear Gil.,

Your old friend and affte. humble servant,

J. Mulso.

This letter ends the correspondence on the side of Mulso, who died in the following September. Like so many of his contemporary clergymen in the eighteenth century, his view of life was not perhaps a very dignified one. Nevertheless, he certainly was a true and attached friend to Gilbert White; and all unknowingly acted the part of a veritable Boswell, by which he has very materially contributed to our knowledge of the Naturalist's career.

To R. Marsham.   Selborne, Jan. 18th, 1791.

Dear Sir,—As your long silence gave me some uneasiness lest it should have been occasioned by indisposition; so the sight of your last obliging letter afforded me much satisfaction in proportion.

I was not a little pleased to find that your friend Lord Suffield corroborated the account of the Cuckoo given by Mr. Jennor, * whose relation of the proceedings of that peculiar bird is very curious, new, and extraordinary. It does not appear from your letter that you endeavoured to revive the Swallow, which fell down before your parlor-window. I have not yet done with trees, and shall therefore add, that my tall 74 ft. beech measures 6 feet in the girth at two feet above the ground. Beeches seem to me to thrive

---

* Jenner, the inventor of vaccination.

best on stoney, or chalkey cliffs, where there seems to be
little or no soil.   Thus about a mile and an half from me to
the S.E. in an abrupt field, stand four noble beech-trees on
the edge of a steep, rocky-ravin, or water-gulley, the biggest
of which measures 9 feet 5 inches at four feet from the
ground.   Their noble branching heads, and smooth rind
show that they are in the highest vigour and preservation.
Again the vast bloated, pollard, hollow beeches, mentioned
before, stood on the bare naked end of a chalky promontory
many of which measured from 20 to 30 feet in circumference! they were the admiration of all strangers.   How
has prevailed the notion that all old London was built with
chestnut?   It is with us now vile timber, porous, shakey,
and fragile, and only fit for the meanest coopery purposes.
Yet have I known it smuggled into Portsmouth dock as
good ship-building oak!

The more I observe and take notice of the best oaks now
remaining in this neighbourhood, the more I am astonished
at the oak which you planted yourself.   For there is a
most noble tree of that kind near Hartely house, which
I caused to be measured last week; when behold, at four
feet above the ground the girth proved to be only 14 feet,
when yours measured 12 ft. 6 in.!   Why this fine shafted
tree, with it's majestic head escaped the axe thirty years ago,
when Sir Simeon Stuart felled all its contemporaries, I
cannot pretend to say.   If you ever happen to see the
Hamadryad of your favourite Oak, pray give my respects
to her.   She must be a fine venerable old lady.   For a
diverting story respecting an Hamadryad, see the 'Spectator,'
vol. viii., p. 128.

Behind my house I have got an outlet of seven acres
laid out in walks by my father.   As the soil is strong, the
hedges, which are cut-up, are prodigious.   The maples about
35 feet in height, and the hasles, and whitethorns 20, which,
when feathered to the ground, were beautiful : but they now,

being 50 years old, have rather over-stood their time; and besides, the severity of December, 1784, has occasioned irreparable damages among the branches. Thus much for trees. Lord Stawell has lately sent me such a bird, sprung and shot in his coverts, as I never saw before, or shall again. I pronounced it to be a mule, bred between a cock pheasant, and a pea-hen.

You say wood-cocks in their passage strike against lighthouses on your coast: a gentleman tells me, that at Penzance sea-fowls frequently dash in the night against windows where they see a light. My well is 63 feet in depth; yet in very dry seasons, as last autumn, it is nearly exhausted: yet you would be surprised to see how few inches of rain falling will replenish it again. How do rains insinuate themselves to such depths? The rains this winter have been prodigious! In November last 7 inches; in December 6 inches. The whole rain at Selborne in 1790 was 32 inches. Sure such thunder, and lightning and winds have never fallen out within your observation in one winter! Had I known You 30 years ago, I should have been much pleased; because I would have gone to have seen you; and perhaps You might have been prevailed on, when all our timber was standing, to have returned the visit. In the year 1746 I lived for six months at Thorney in the Isle of Ely, to settle an executorship, and dispose of live stock: there I lost nine oxen by their eating yew, as mentioned in my book. I hope you will write not long hence. With the truest respect and esteem I remain,

Your most humble servant,

GIL. WHITE.

The dark butterfly which you saw was the *papilio urticæ*: it is often more early than the yellow *papilio rhamni*. At this moment the Barometer stands somewhat below 28·5 in.; the rain this day has been very great from the S.E.!

To R. Marsham.

Selborne, Feb. 25th, 1791.

Dear Sir,—It was elegantly remarked on our common friend, and my quondam neighbour Doctor Stephen Hales, by one who has written his character in Latin, that—"scientiam philosophicam usibus humanis *famulari* jussit." The observation was just, and the assertion no inconsiderable compliment : for undoubtedly speculative enquiries can bear no competition with practical ones, where the latter profess never to lose sight of utility.

As I perceive You loved the good old man, I do not know how I can amuse You better than by sending you the following anecdotes respecting him, some of which may not have fallen within your observation. His attention to the inside of Ladies' tea-kettles, to observe how far they were incrusted with stone (*tophus lebetinus* Linnæi) that from thence he might judge of the salubrity of the water of their wells— his advising water to be showered down suspicious wells from the nozle of a garden watering-pot in order to discharge damps, before men ventured to descend; his directing air-holes to be left in the out-walls of ground-rooms, to prevent the rotting of floors and joists; his earnest dissuasive to young people, not to drink their tea scalding hot; his advice to water-men at a ferry, how they might best preserve and keep sound the bottoms or floors of their boats; his teaching the house-wife to place an inverted tea-cup at the bottom of her pies and tarts to prevent the syrop from boiling over, and to preserve the juice; his many though unsuccessful attempts to find an adequate succedaneum for yeast or barm, so difficult to be procured in severe winters, and in many lonely situations; his endeavour to destroy insects on wall fruit-trees by quick-silver poured into holes bored in their stems; and his experiments to dissolve the stone in human bodies, by, as

I think, the juice of onions;—are a few, among many, of those benevolent and useful pursuits on which his mind was constantly bent. Though a man of a Baronet's family, and of one of the best houses in Kent, yet was his humility so prevalent, that he did not disdain the lowest offices, provided they tended to the good of his fellow creatures. The last act of benevolence in which I saw him employed was, at his rectory of Faringdon, the next parish to this, where I found him in the street with his paint-pot before him, and much busied in painting white, with his own hands, the tops of the foot-path posts, that his neighbours might not be injured by running against them in the dark. His whole mind seemed replete with experiment, which of course gave a tincture, and turn to his conversation, often somewhat peculiar, but always interesting. He used to lament to my Father, how tedious a task it was to convince men, that sweet air was better than foul, alluding to his ventilators: and once told him, with some degree of emotion, that the first time he went on board a ship in harbour at Portsmouth, the officers were rude to him; and that he verily believed he should never have prevailed to have seen his ventilators in use in the royal navy, had not Lord Sandwich, then first Lord of the Admiralty, abetted his pursuits in a liberal manner, and sent him down to the Commissioners of the dock with letters of recommendation. It should not be forgotten that our friend, under the patronage of Sir Joseph Jekyll, was instrumental in procuring the Gin-act, and stopping that profusion of spirituous liquors which threatened to ruin the morals and constitutions of our common people at once. He used to say, that the hogs of distillers were more brutal than the hogs of other men; and that, when drunk, they used to bite pieces out of each other's backs and sides! With due respects I remain

Your most humble servant,

GIL. WHITE.

I did myself the honour of writing to you very lately about trees, and other matters. This winter continues wet and mild: wet springs are bad for Selborne. My crocus's make a fine show.

During the spring and early summer of this year many relations were entertained at Selborne. On June 23rd their host visited Mr. Edmund Woods at Godalming; and records that he "went to see the village of Compton, where my father lived more than sixty years ago, and where seven of his children were born. The people of the village remember nothing of our family." The 7th of July found him at South Lambeth, when he noted the contents of "a fruit-shop near S. James," viz. "black cluster grapes, pine apples, nectarines, and Orleans plums."

"July 12th. On this day my Bro. Benjn. White began to rebuild his house in Fleet Street, which he had entirely pulled to the ground. His grandson, Ben. White, laid the 1st brick of the new foundation, and then presented the workmen with five shillings for drink. Ben., who is 5 years old, may probably remember the circumstance hereafter, and may be able to recite to his grandchildren the occurrences of the day."

So man proposes. This Benjamin White died young, and the publishing business did not remain long in the White family.

The following letter affords the first evidence of the writer's really serious illness :—

Benjamin White

Walker & Cockerell, ph. sc.

To Benjn. White.   Selborne, July 23rd, 1791.

Dear Brother,—It is full time to make our proper acknowledgements for all your favours, and good offices which we
experienced at your house for so long a time; and especially
as we are assured that you will be pleased to hear that we
got safe to our journey's end. My gravel, I thank God,
did not return in consequence of my journey, and my
feverish disorder has partly left me: so that I hope to have,
for a time, a respite from my sufferings. The showers of
Monday 18th availed us as far as Wimbledon-common; but
afterwards the dust became very troublesome. When we
came to this place we found that this hill-country had lately
enjoyed many refreshing showers; and that the gardens and
fields were much benefited thereby; though the former
months had been so very hot and dry, that at an average
the farmers did not get more than half a crop of hay. I
am ashamed to tell you what a blunder I made at Cobham
by paying the Clapham driver *four shillings* more than his
due. This fellow pothered me, by insisting that his master
was to have 18*s.* instead of 17*s.* and would have perswaded
me that I always paid 18*s.* from Cobham to S. Lambeth.
At length I laid down a guinea, and reckoning from 17*s.*
paid him five more to make 22*s.*, which was due for the
journey, and his attendance the Friday before. So that
I paid him 26*s.* instead of 22*s.*; besides two shillings which
I *gave* him as driver. The money that I *paid* him was one
guinea and two half-crowns, as he must well remember. Of
this mistake I was not conscious till we were some way out
of Cobham.

Mrs. Edmund White has advanced my nephews and nieces
to the number of 58.

                Y^r obliged and affectionate brother.

Wheat looks well, and the Hops at this place are in a
promising way. I return Mr. B. White many thanks for
his letter of July 21st.

From the *Naturalist's Journal*—

"[1791] July 30." [Mrs.] Ben. White writes that "my
father shot in his own garden at S. Lambeth a *Loxia
curvirostra*, or Crossbill, as it was feeding on the cones of
his Scotch firs. There were six, four cocks and two hens."

To B. White, senr.   Selborne, Dec. 8, 1791.

Dear Brother,—I am to thank you for the tin gutter
stove pipe &c. which seem to have come safe: but I have
as yet employed no workman to put them up, or examine
them.

Be pleased to show this to my nephew Ben. White and he
will be so kind as to repay you £1 12s. 0d. laid out for the
spouts above mentioned which, I trust, will save my front
wall. I have half a mind to build me a good light closet
to my bedchamber and under it a kind of pantry, and store
room for the kitchen; both of which would much improve
my house: but whether at my time of life I shall have
resolution to set about such a jobb, I much doubt.

Dr Chandler talks, some times, as if he should not continue a great while longer at this place; now should he
remove some time hence, you might, I should suppose, have
the choice of a *roomy* old house and some acres of land,
where you might amuse yourself till something better
offered. You would then become our neighbour indeed.

You have, I find, paid the eight guineas for the mare,
which will, I hope, prove a good bargain, and perform a
good deal of moderate work. Poor Marlow was buried
this evening: Tanner still keeps in bed.

We are glad to hear that your house in town goes on
so well. The boys are well.

Mrs. J. White joins in respects.

Y^r loving brother,
GIL. WHITE.

Rain in Oct. 6.49: in Novr. 8.16! From the 13th of Novr. to the 19th both inclusive, viz. in one week, the rain was 5.10!! There was thunder two nights. I have just fixed a white cross, and capped the Hermitage with a new coat of thatch, so that to us below it becomes a very picturesque object. Hale's eldest daughter has eloped from her wretched husband; and went, it is said, with a married man, who has children grown men and women.

The Hermitage mentioned in the last letter was, no doubt, the *new* Hermitage, situated close to the Bostal, immediately opposite Gilbert White's house. A small, but yet perfect, zigzag path through the trees leads up to where it stood from a wicket gate in the little park, which looks as though it might have existed in the Naturalist's time.

To R. Marsham.   Selborne, near Alton,

Decr. 19, 1791.

Dear Sir,—Your letter,* which met me so punctually in London, was so intelligent, and so entertaining, as to have merited a better treatment, and not to have been permitted to have lain so long unnoticed!

That there is no rule without an exception is an observation that holds good in Natural History: for though you and I have often remarked that Swifts leave us in general by the first week in August: yet I see by my journal of this year, that a relation of mine had under the eaves of his dwelling house in a nest a young squab swift, which the dam attended with great assiduity till September 6th, and

---

* Of July 8th, 1791. It commences, "My thanks are greatly due to you for the favour of your pleasing letter of the 8th of June." This letter is unfortunately missing, though possibly it still exists in some collection of autographs.

on October 22nd, I discovered here at Selborne three *young martins* in a nest, which the dams fed and attended with great affection on to November 1st, a severe frosty day; when they disappeared; and one was found dead in a neighbour's garden. The middle of last September was a sweet season! during this lovely weather the congregating flocks of house martins on the Church and tower were very beautiful and amusing! When they flew off all together from the roof, on any alarm, they quite swarmed in the air. But they soon settled again in heaps on the shingles; where preening their feathers, and lifting up their wings to admit the rays of the sun, they seemed highly to enjoy the warm situation. Thus did they spend the heat of the day, preparing for their Migration, and as it were consulting when and where they are to go! The flight about the church consisted chiefly of house martins, about 400 in number: but there were other places of rendezvous about the village, frequented at the same time. The swallows seem to delight more in holding their assemblies on trees. Such sights as these fill me with enthusiasm! and make me cry out involuntarily,

> "Amusive birds! say where your hid retreat,
> When the frost rages, and the tempests beat!"*

We have very great oaks here also on absolute sand. For over Wolmer forest, at Bramshot place where I visit, I measured last summer three great hollow oaks, which made a very grotesque appearance at the entrance of the avenue, and found the largest 21 feet in girth at five feet from the ground. The largest Sycamore in my friend's court measures 13 feet. His edible chestnuts grow amazingly, but make (for some have been felled) vile *shaky, cupshaky* timber. I think the oak on sands is shaky, as it is also

---

* A quotation from 'The Naturalist's Summer Evening Walk,' appended to Letter XXIV. to Pennant.

on our rocks, as I know by sad experience the last time I built. The indented oaken leaf which you gathered between Rome and Naples was the *quercus cerris* of Linnæus. The yellow oak which you saw in Sussex escaped my notice.

Richard Muliman Trench Chiswell, Esq., of Portland Place, and M.P., tells a friend of mine in town that he has an *Elm* in Essex for which he has been bid £100. It is long enough, he says, to make a keel ungrafted for a man-of-war of the largest dimensions. As he expressed a desire of corresponding with me, I have written to him, and desired some particulars respecting this amazing tree.

You seem to wonder that Willughby should not be aware that the Fern-owl is a summer bird of passage. But you must remember that those excellent men, Willughby and Ray wrote when the ornithology of England, and indeed the Natural History, was quite in its infancy. But their efforts were prodigious; and indeed they were the Fathers of that delightful study in this kingdom. I have thoughts of sending a paper to the R. S. respecting the fern-owl; and seem to think that I can advance some particulars concerning that peculiar migratory, nocturnal bird, that have never been noticed before. The rain of October last was great, but of November still more. The former month produced 6 in. 49 hund., but the latter upwards of 8 in.: five and $\frac{1}{2}$ of which fell in one week, viz. from Nov. 13th to the 19th both inclusive! You will, I hope, pardon my neglect, and write soon. O, that I had known you forty years ago!

> I remain, with great esteem,
>> Your most humble servant,
>>> GIL. WHITE.

My tortoise was very backward this year in preparing his Hybernaculum; and did not retire till towards the beginning of December. The late great snow hardly reached us, and was gone at once.

238 GILBERT WHITE OF SELBORNE 1792

Mr. Churton arrived on December 23rd to spend his usual Christmas visit at Selborne, leaving again early in January.

To Miss Mary Barker.   Feb. 18th, 1792.

Rain at Selborne in 1791.

Jan.	.	.	. 673
Feb.	.	.	. 464
Ma.	.	.	. 159
Apr.	.	.	. 113
May	.	.	. 133
Ju.	.	.	. 91
Jul.	.	.	. 556
Aug.	.	.	. 173
Sep.	.	.	. 173
Oct.	.	.	. 649
Nov.	.	.	. 816
Dec.	.	.	. 493

44.93

Dear Niece,—I herewith send you an account of the last year's rain, which was very great, and in particular in November, when there fell from the 13th to the 19th both inclusive about 510. We were surprized to hear of the vast snows, and severe weather that you experienced in December because all the while we had little snow, and no frost of any continuance. We condole with you on the loss of old Mrs. Barker; who yet seems to have been a happy woman: for after a blameless life, spent in affluence and comfort among affectionate relations, she departed this life in peace at the good old age of 90 and upwards. I have disposed of your mother's guinea with much satisfaction among such old people as seemed to want it most. Old Dewye, and wife, are alive, but almost childish; and old Geo. Tanner;

but he has been confined to his bed for three months. Charles Etty did not come home in his own ship (in which he went out second mate) because it was said that he broke his leg at Madras the very evening before the ship was to have sailed.   Poor dear Caroline Bingham was a most amiable girl and a fine figure: but she dyed suddenly as soon as she left this place, to the great sorrow of her parents! they have several more children.   Dr Chandler is in London settling the concerns of his brother; he was a clergyman in Surrey, and has left a daughter grown-up. Mrs. Chandler looks a little as if she was intending to encrease her family not long hence.   The death of my good friend Mr. Mulso is a sad loss to his children: where his daughters are to live we have not heard.   My brother Benjamin, we hear, begins to think seriously of relinquishing his business to his sons; and meditates a retreat into Hants for the remainder of his life, intending to leave S. Lambeth. Perhaps he may settle at Marelands, a beautiful seat between Alton and Farnham, late the residence of Mr. Sainesbury, Uncle to Mrs. Edmund White, and Agent to Lord Stawell, Ld Salisbury, the Marquis of Downshire, &c., &c.   This gentleman dropped suddenly out of his chair, and was dead in a moment, on the eve of his birthday, while his wife was preparing an elegant entertainment for his friends the day following.   Mr. S. was a man of an excellent character, and beloved by every body.   Mr. Clement, very fortunately, is to succeed his friend in his agencies for Lord Stawell and Mrs. Beckford: these employs will make a very handsome addition to Mr. Clement's income, and will give him credit and reputation in this neighbourhood.   Mrs. J. White desires to join in best respects to all your family, and to friends at Whitwell and Stamford.

I remain

Y^r loving uncle,

GIL. WHITE.

We have enjoyed lately sweet summer weather: but last night a most severe frost came on, with snow, and Thermometer at 21°! Newton friends lay here last night.

Marelands house and farm belong to Ld Stawell.

To R. Marsham. Selborne, Mar. 20th, 1792.

Dear Sir,—You, in a mild way, complain a little of *Procrastination*: but I, who have suffered all my life long by that evil power, call her the *Dæmon of Procrastination*; and wish that *Fuseli*, the grotesque painter in London, who excells in drawing witches, dæmons, incubus's, and incantations, was employed in delineating this ugly hag, which fascinates in some measure the most determined and resolute of men.

You do not, I find, seem to assent to my story respecting Mr. Chiswell's elm. There may be probably some misapprehension on my side. I will therefore allow Mr. Ch. that privilege which every Englishman demands as his right, the liberty of speaking for himself. "In regard to my tree," says he, "it is a *Wych Elm*, perfectly strait, and fit for the keel of the largest man of war. The purveyor of the navy offered my late Uncle £50 for it, although it would have cost as much more to have conveyed it to Portsmouth; and he would have run all risque of soundness. It grows about eleven miles from Safron Walden, in a deep soil, and near 30 from Cambridge, the nearest place for water-carriage. I will measure it next summer." He adds, "I have been, and am a considerable planter; and have been honoured with three gold medals from the Society of Arts," &c. Thus far Mr. Ch.

As I begin to look upon You as a Selborne man, at least as one somewhat interested in the concerns of this place; I wish that you could see 'The sixth Report of the Commissioners appointed to enquire into the state, and condition of the Woods, Forests, &c. of the Crown,' &c. This Report was printed February, 1790; though never published: but distributed among the members of the house of commons from some of whom you may borrow it, as I have done. This curious survey will inform you, from the best authority, of all the circumstances respecting the advantages, usages, abuses, &c. of our Forest of Alice Holt, and Wolmer. Here you will see, that the Forest now consists of 8,694 acres, 107 of which are in ponds; that the present timber is estimated at £60,000; that it is almost all of a size, and about 100 years old; that it is shamefully abused by the neighbouring poor, who lop it, and top it as they please; that there is no succession because all the bushes are destroyed by the commoners around; that your old favourite Oak, the *Grindstone Oak*, is estimated at 27 loads of timber; that the peat cut in Wolmer is prodigious; in the year 1788 in one walk 942 loads; and in another walk the same year 423 loads, besides heath and fern; and in the same year 935,000 turves; &c. &c. &c. Lord Stawell is the Lieutenant, or Grantee, whose lease expires in 1811, as I have said in my book. That Nobleman did me the honour to call on me a morning or two ago, and sat with me two hours: he brought me a white wood cock, milk white all over except a few spots.

My friend at Bramshot place,* where I measured the great pollard oaks, and Sycamore last summer, has got a great range of chestnut-paling; I shall tell him what Mr. Kent says respecting timber of that sort. The rain with us in 1791 was 44 in. 93 hund.: upwards of 8 inches of which fell in November! the rain of the present year has been considerable. Our indications of spring this year are thus: Jan. 19 *winter-aconite* blows: Jan. 21 *Hepaticas* blow. Jan. 29. *snowdrop* blows: 31 Hasels: Feb. 4 *Crocus* b. 13. *brimstone butterfly*; 21. *yellow wagtail* appears. 26. *Humble*

---

* Near Liphook. The friend was a Mr. Richardson, whom he sometimes visited for a day or two.

bee: March 16. *daffodil blows,* and *Apricot*: 19. *peaches,* and *nectarines.* I have read *Boswell's Johnson* with pleasure. As to Bishop Horne I knew him well for near 40 years : he has often been at my House. Stillingfleet, I see, wrote his *Calendar of Flora* at your house : He speaks in high terms of the hospitable treatment that he experienced at Stratton.

Wonderful is the regularity observed by nature! I have often remarked that the smallest *willow wren* (see my Book), called here the *Chif-chaf* from its two loud sharp notes, is always the *first spring bird* of passage, and that it is heard usually on March 20 : when behold, as I was writing this very page, my servant looked in at the parlour door, and said that a neighbour had heard the *Chif-chaf* this morning!! These are incidents that must make the most indifferent look on the works of the Creation with wonder !

My old tortoise lies under my laurel hedge, and seems as yet to be sunk in profound slumbers. You surprise me, when You mention your age : your neat hand, and accurate language would make one suppose you were not 50.

I remain, with true esteem,

Y^r most obliged servant,

GIL. WHITE.

When Mr. Townsend avers that the Nightingales at *Valez* sing the winter through, I should conclude that he took up that notion on meer report; because I had a brother who lived 18 years at Gibraltar, and who has written an accurate Nat. Hist. of that rock, and its environs. Now he says, that Nightingales leave Andalusia as regularly towards autumn as other Summer birds of passage. A pair always breeds in the Governor's garden at the Convent. This History has never been published, and probably now never will, because the poor author has been dead some years. There is in his journal such ocular demonstration of *swallow emigration* to and from Barbary at Spring, and fall, as, I know, would delight you much. There is an *Hirundo hiberna*, that comes to Gibraltar in October and departs in March; and abounds in and about the garrison the winter thro'.

On April 4th Gilbert White went to Oxford. All his life he had had little reason to complain of his health, which appears to have been excellent; and there is nothing to show that he had any reason to suppose that, when he returned to Selborne on April 17th, he had said good-bye to the common-room at Oriel for the last time. Such was, however, the case, since in the following spring he was not well enough to leave home for Oxford.

In May, 1792, he received a visit from his niece, Mrs. B. White, junr., and her son Tom, and in July the Provost of Oriel and Mrs. Eveleigh came. Mr. Churton came again at this time.

To R. Marsham.   Selborne, Augst. 7, 1792.

Dear Sir,—While all the young people of this neighbourhood are gone madding this morning to the great last day's review at Bagshot; I am sitting soberly down to write to my friend in Norfolk; almost forgetting, now I am old, the impulse that young men feel to run after new sights; and that I myself, in the year 1756, set-off with a party at two o' the clock in the morning to see the Hessian troops reviewed on a down near Winchester.* While I was writing the sentence above, my servant, and some neighbours came down from the hill, and told me that they could not only hear the discharges of the ordnance and small arms, and see

---

* This journey was made on October 5th and 6th, 1756, with his brother Benjamin.

the volumes of smoke from the guns; but that they could also, they thought, smell the scent of the gun-powder, the wind being N.E. and blowing directly from the scene of action at Wickham bushes, tho' they are in a direct line more than twenty miles from hence.

As I had written to you as long ago as March, I began to fear that our correspondence was interrupted by indisposition; when your agreeable letter of July 14th came in, and relieved me from my suspense. You do me much honour by calling one of your beeches after my name. Linnæus himself was complimented with the *Linnæa borealis* by one of his friends, a mean, trailing, humble plant, growing in the steril, mossy, shady wilds of Siberia, Sweden, and Russia; while I am dignified by the title of a stately Beech, the most beautiful, and ornamental of all forest trees. The reason, I should suppose, why your trees have not encreased in growth, and girth this summer is the want of heat to expand them. I have not this year measured my firs in circumference, but they have, I see, many of them, made surprising leading shoots. My account of the *Fern-owl*, or *Eve-jarr* was prevented by *Madam Procrastination*, who, a jade, lulled me in security all the spring, and told me I had time enough, and to spare, till at last I found that the R. S. meetings were prorogued till the autumn; against which I hope to be ready: and as I have got my materials, trust that when I do set about the business "verba haud invista sequentur." By *all means* get a sight of the *sixth Report* of the *Commissioners*, &c., it will entertain you, and furnish you with much matter, and many anecdotes respecting Selborne, of which I could have availed myself greatly had they been printed before I published my work. My book is gone to Madras, and several to France, and one to Switzerland, and one copy is going to China with Lord Macartney: but whether some Mandareen will read it, I know not. We have a young

gentleman here now on a visit, the son of our late Vicar
Etty, who assures me, that at Canton he has seen the
Chinese reading English books; and has heard them converse sensibly on the manners, and police of this kingdom.
The *Chif-Chaf* of this village is the smallest *willow-wren*
of my History. Once I had a spaniel that was pupped
in a rabbit burrough on the verge of Wolmer forest.
Though I have long ceased to be a sportsman, yet I still
love a dog; and am attended daily by a beautiful spaniel
with long ears, and a spotted nose and legs, who amuses
me in my walks by sometimes springing a pheasant, or
partridge, and seldom by flushing a woodcock, of late
become with us a very rare bird. Remember the story
of Pylades and Orestes; and do not say that exalted friendship never existed among men.*   Chif-Chaf, the first bird
of passage, was heard here March 20th: *Swallow* was seen
March 26th: *Nightingale,* and *Cuckoo* April 9th: *House-
martins* April 12th: *Redstart* April 19th: *Swift* April 14th:
Fern-owl heard May 19th: *Fly-catcher,* the latest summer
bird, May 20th. We have experienced a very black wet
summer, and solstice; but none of those floods and devastations mentioned in the newspapers! Indeed we know no
floods here, but frequent rains. Yet in warm summers we
have as fine melons, and grapes, and wall-fruit as I have
ever seen. July at an average produces the most rain of
any English month. This last measured 5 in. and 15 h.
Pray, good Sir, procure better ink; your's is so pale, that
it often renders your neat hand scarcely legible! I am now
offering my intelligent young neighbours *sixpence* for every
authentic anecdote that they can bring me respecting *Fern-
owls,* and will give you the same *sum* for the same information. As I was coming over our down after sun-set lately,
a cock bird amused us much by flying round and settling

---

* In his last letter Marsham had told a touching story of the affection of
one dog for another.

often on the turf. As he passed us, he often gave a short squeak, or rather whistle. We were near his nest. These, like other birds of passage, frequent the same spots. There are always three pairs on our hill every year. Did you know Sir John Cullum of your part of the world? He was an agreeable, worthy man, and a good antiquary. I was also well acquainted with your late good Bishop Horne: he has often been at my house. I concur with you most heartily in your admiration of the harmony and beauty of the works of the creation! Physico-theology is a noble study, worthy the attention of the wisest man! Pray write. Our swifts have behaved strangely this summer: for the most part there were but three round the church, except now and then of a fine evening, when there were 13. They seem to be all gone. House-martins leave Gibraltar by the end of July. I conclude with all due regard.

Y^r Humble Servant,

GIL. WHITE.

To R. Marsham.

Selborne, November 3, 1792.

Dear Sir,—An extract from the Natural History of Gibraltar by the late Reverend John White.

"In the first year of my residence at Gibraltar which was 1756, it appeared extraordinary to me to see birds of the *Swallow* kind very frequent in the streets all the winter through. Upon enquiry I was told that they were *Bank Martins*: and having at that time been but little conversant in Natural History, they passed with me as such for some years without any farther regard. At length, when I had taken a more attentive survey of the physical productions of this climate, I soon discovered these birds to be none of the common *British* species described by authors; and I farther found that they were never seen in Gibraltar through the whole course of the *summer*; but constantly and invariably made their first appearance about the 18th and 20th, and once as early as the 12th of October and remained in great abundance until the beginning of *March*.

"These phænomena awakened and alarmed my curiosity as events entirely new and unheard of among the body of Ornithologists, and induced me to be particularly exact and attentive in my observations on every part of their conduct. Early in the autumn vast multitudes of these martins congregate in all parts of the town of *Castillar*, which is situate on the summit of a precipice most singularly lofty and romantic, about 20 miles north of *Gibraltar*. Hence it may be inferred that they build and breed on the inland mountains of *Andalusia* and *Grenada*. But on the approach of winter, when their summer habitations become bleak and inhospitable, (for all those mountains are then usually covered with snow) they retreat to these warm shores, and remain there till the snow is gone next spring. A few are always to be seen about our hill by the middle of October shifting round to all sides of the rock at times to avoid the wind. November 2nd, 1771, I saw several, with some young ones among them sitting in groupes, on the cliffs, where the old ones came and fed them."

Thus have I, for your amusement, according to promise, sent you an extract concerning this new, and unnoticed *swallow*, which my Brother, with great propriety, in his work has called *Hirundo hyemalis*; and has given several particulars concerning it, and a description of it, too long for the compass of a letter.

Permit me just to hint to you, that I wrote to you some time ago in answer to your last letter, which gave me much satisfaction.

I forgot to mention in the extract, that these *winter Swallows* usually leave Gibraltar about the beginning of March, unless deep snows (as is sometimes the case, and was particularly so in 1770 and 1772) fall in Spain about that time; and then they linger there till the latter end of the month.

Surely my dear Sir, we live in a very eventful time, that must cut-out much work for Historians and Biographers! but whether all these strange commotions will turn out to the benefit or disadvantage of old England, God only knows! We have experienced a sad spring, summer, and autumn: and now the fallows are so wet, and the land-springs break forth so frequently, that men cannot sow their wheat in any comfort. Our barley is much damaged; and malt will be bad.

Have you read Mr. Arthur Young's 'Travels through France'? He says (p. 543), when speaking of the French clergy—" One did not find among them poachers, or fox-hunters who having spent the morning in scampering after hounds, dedicate the evening to the bottle, and reel from inebriety to the pulpit." Now, pray, who is Mr. Young; is he a man of fortune, or one that writes for a livelihood? He seems to reside in Suffolk, near Bury St. Edmund; so probably you can tell me somewhat about him.

Pray do *wood-peckers* ever damage, and bore your timber-trees? not those, I imagine, of your own planting, but only those that are tending to decay. I had a brood this year in my outlet hatched, I suspect, in the bodies of some old willows. My dissertation on the *Caprimulgus* is almost finished.

I remain, with all due respect, and esteem,
Your most obedient and obliged servant,
GIL. WHITE.

On November 10th Benjamin White quitted South Lambeth, and came to reside at his house Mareland, near Farnham.

To the Rev. R. Churton.

Selborne, Nov. 15, 1792.

Dear Sir,—As your own account of the bad state of your health, written to Dr Chandler, gave us much concern, so in proportion your late cheerful letter to Mrs. Chandler afforded us no small satisfaction. I sit down now to invite you to spend part of your Xtmass holidays with us. But as your usual time of vacation, when divided into two parts, will be little or nothing, we hope you will be able to extend your furlow. You have of late years paid me a compliment for varying my phrases of invitation; but all those terms of words are exhausted, and I have now nothing left but the plain, honest assertion of wishing to see you, as often and as long as you can make it agreeable and convenient to yourself.

I return you my best thanks for your quotation from Aristotle, of which I hope to avail myself soon; and for a *correct* copy of the inscription on the tomb of the great Mr. Ray. It is pleasant to hear that friends to Genius are still to be found, who, at periods, are ready to repair and beautify the monument of departed worth, nor suffering it to be effaced with weeds and filth. However his *works* will be, as the inscription says, the most lasting monument of his fame. Every time you come, I have been provided with a new book for your inspection. In some respects you will think Mr. Arthur Young's 'Journey in France' reprehensible; and will not always subscribe to his politics. However the writer is a man of observation, and has a curious chapter on Climate. In three summers he threaded every corner of that vast kingdom, and made an excursion through the Pyrenees to Barcelona, and another over the Alps and Apennine to Turin, Venice, Florence, &c. Mr. Young, I fear, is no friend to us parsons. Mr. Marsham has just sent me a long letter; but he complains of infirmities.

Mrs. J. White joins in good wishes; and desires respects to the Provost, when you see him; and to the Cox family, Dr Nowel, &c. &c.   With all due regard I remain,

Yours affectionately,

GIL. WHITE.

Take care of your health, and don't study too hard. When the shell of your House is compleat, insure it. A friend of mine at Salisbury has just had a house, not quite finished, burnt to the ground.   It was to have cost £4,000!

To R. Marsham.   Selborne, Novr. 20, 1792.

Dear Sir,—Our last two letters seem as if they had crossed each other on the road; but whether they conversed when they met does not appear.

If you have got the *Certhia muraria*, or true *Wall-creeper*, you are in possession of a very rare and curious bird. For in all my researches here at home for 50 years past, and in all the vast collections that I have seen in London I have never met with it. No wonder that the great Mr. Willughby is not very copious on the subject, for he acknowledges fairly that he had not seen it; though he supposes it may be found in this island. The best person I can refer you to is, Dr *John Antony Scopoli*, a modern, elegant, foreign Naturalist, born in the *Tyrol*, but late deceased in Pavia, where he was professor of Botany. This curious, and accurate writer was in possession of one in his own Museum, and gives the following description of his specimen in his '*Annus primus historico-naturalis*': "that its bill is somewhat longer than its shanks, slender, and somewhat bent; that the tongue is bifid; and the feet consisting of three toes forward and one behind." Again he adds, "that the upper part is cinerous, the throat whitish; the abdomen, wings in part, tail and feet, black: the wings at their base, and the quill feathers at their base on one side reddish." "It was taken in Carniola." "It is the size of the common *Creeper*,* or *Certhia familiaris*: its nostrils oblong; tail cinerous at the point; the first four quill feathers distinguished on the inner side by two white spots." He concludes thus,— "Migrat solitario sub finem autumni; turres et muros œdium altiorum adit; araneas venatur; saltitando scandit; volatu vago et incerto fertur volucris muta." You are sure, I trust, that your bird is not the *Sitta Europœa*, or *Nut-hatch*.

I have written so soon, that you may examine your bird well again, before the specimen decays. Your Lady's turkey-hen is a most prolific dame; and must, I think, lay herself to death. You persist, very laudably in your curious experiments on trees. Whenever you recommend my book, which begins to be better known, you lay me under fresh obligations. I am writing my account of the Fern-owl, and endeavouring to vindicate it from the foul imputation of being a *Caprimulgus*. My letter will make a fierce appearance with a quotation from *Aristotle*, and another from *Pliny*: but whether the R. S. will read it: or whether afterwards they will print it, I know not.

With all good wishes for your health, and prosperity I remain

Your obliged, & humble servant,

GIL. WHITE.

The history of the Fern-owl was, however, never completed. Nothing shows the Selborne Naturalist's general knowledge of ornithology better than the way in which he recognised from description the Wall Creeper shot at Stratton, though he had never seen the bird.

---

* This is a slip of White's pen. Scopoli's words (*op. cit.*, p. 51) are, 'Statura *sittæ*,' that is, the size of the Nut-hatch, which is nearly true.—A. N.

CHAPTER XI.

To Mrs. Barker.

Selborne, Jan. 2, 1793.

Dear Sister,—While Mrs. J. White is employed in knitting, and Mr. Churton in reading and writing, I sit down, as I have usually done at this season of the year, to send Mr. Barker the quantity of rain, and you some account of our welfare. Ned White,* you may have heard, is settled with a banker in London, where he gives satisfaction, and is allowed £50 per ann. Gil. White* has been so unfortunate as to lose his Master, an attorney at Bath, by death, after he had served 3 years; and what was worse, the man dyed insolvent. By this untoward accident the poor young man has been thrown out of employ for three or four months: but, by the interest of friends, was reinstated in business yesterday with a gentleman at Petersfield,† where he is to stay three years more without premium; but must pay for his board. The first premium, £200, is all lost! Mr. and Mrs. B. White have lately been with us for a few days, and both seemed very well. Poor Nanny Woods's new husband is in a dangerous decline. Much used to be said of his bad health for some time past; and therefore it is a pity that the match took place! Dr Chandler keeps improving the parsonage-house, and therefore, I conclude, has

---

* Sons of Henry White.
† His uncle Gilbert giving £50 towards this arrangement.

252

no thoughts of moving. He has taken off an entry from the hall, and has made the rest of that room into a good parlor. Much was the damage that we sustained by the late sad wet summer and autumn in our hay, our fallows, our corn, and our forest fuel, which lies rotting in the moors of Wolmer. Our brick-burner, after he had payed duty for a large cargo of bricks and tiles, never could get them dry enough for burning. My fruit never ripened, and especially my grapes. The year 1782, part of which you spent here, was some how less distressing, though the rain was then upward of 50 in., as you may see by my book. Your grandson, I hope, will thrive, and become as honest and good a man as his grandfather and father. Mrs. J. White thanks you for late kind present.* . . . . . add, that there is a probability [that her son?] may soon be married : * . . . . . . fortune remain unsettled with the father, matters at present remain uncertain. Should this match take place, it has the appearance of a very respectable connection. The family lives near Sarum,† and a coach is kept : and the lady, it is said, and her sisters are accomplished, and musical. Mr. Churton was lately presented by Brazenose Coll. to one of their best livings, the rectory of Middleton Cheney in Northamptonshire, but near Banbury, which he hopes will neat him £400 per ann. He is obliged to rebuild part of the house. Mr. Churton joins with us in all the good wishes of the season.

I remain,

Yours affectionately,

GIL. WHITE.

Old G. Tanner is still in bed : yesterday the widow of James Carpenter was buried aged 93. I will bestow your charity in a proper manner, and return you thanks for it.

---

    * Letter imperfect.   † At Downton.

To R. Marsham.

Selborne, Jan. 2, 1793.

Rain in 1792.

	Inch.		Hund.
Jan.	. . 6	...	7
Feb.	. . 1	...	68
Mar.	. . 6	...	70
Apr.	. . 4	...	8
May	. . 3	...	0
Ju.	. . 2	...	78
July	. . 5	...	16
Aug.	. . 4	...	25
Sep.	. . 5	...	53
Oct.	. . 5	...	55
Nov.	. . 1	...	65
Dec.	. . 2	...	11
	―		―
	48	...	56

Dear Sir, — My best thanks are due for your kind letter of December 21st, to which I shall pay proper attention presently. But I shall first speak of the margin of this, which contains the rain of last year, which was so remarkably wet, that you may be perhaps glad to see what proportion the fall of water bears to that of other uncomfortable, unkindly years. The rain in 1782, as you see in my book, was 52 inches; in 1789, 42 inches; and in 1791, 44 inches : yet these wet seasons had not the bad influence of last year, which much injured our harvest; damaged our fallows; prevented the poor from getting their peat and turf, which lies rotting in the Forest; washed and soaked my cleft beechen wood, so that it will not burn; it prevented our fruits from ripening. The truth is, we have had as wet years, but more intervals of warmth and sunshine.

I am now persuaded that your bird is a great curiosity, the very *Certhia muralis,* or *Wall-creeper,* which neither Willughby nor Ray ever saw; nor have I, in 50 years attention to the winged creation, ever met with it either wild, or among the vast collections that I have examined in London. It seems to be a South Europe bird, frequenting towns and towers and castles : but has been found, but very seldom indeed, in England. So that you will have the satisfaction of introducing a new bird of which future Ornithologists will say, " found at Stratton in Norfolk by that painful, and accurate Naturalist, Robert Marsham, Esq." You observe that Scopoli does not take notice that the hind-claw is about

double the length of the fore-claws: but Linnæus corrobo-
rates your remark by saying "Ungues validi, praesertim
posticus." You seem a little to misunderstand Scopoli respecting the spots on the *inner side* of the quill feathers:
by the *inner side* he does not mean the *under side* of the
wing next the body; but only the *inner* or *broader web* of
the quills, on which those remarkable spots are found, as
appear by the drawing. I am much delighted with the
exact copies sent me in the frank, and so charmingly
executed by the fair unknown, whose soft hand has directed
her pencil in a most elegant manner, and given the specimens a truly delicate, and feathery appearance. Had she
condescended to have drawn the whole bird, I should have
been doubly gratified! It is natural to young ladies to wish
to captivate men: but she will smile to find that her present
conquest is a very old man.

My best thanks are due for all your good offices respecting
my work, and in particular for your late recommendation to
the Duke of Portland.

You did not in your last, take any notice of my enquiries
concerning *wood-peckers*, whether they ever pierce a sound
tree, or only those that are tending to decay. I have
observed that with us they love to bore the edible chestnuts; perhaps because the wood is softer than that of oak.
They breed in my outlet, I think in old willows. You have
not told me anything about Arthur Young. You cannot
abhor the dangerous doctrines of levellers and republicans
more than I do! I was born and bred a Gentleman, and
hope I shall be allowed to die such. The reason you have
so many bad neighbours is your nearness to a great factious
manufacturing town. Our common people are more simple-
minded and know nothing of Jacobin clubs.

I admire your fortitude, and resolution; and wonder that
you have the spirit to engage in new woods, and plantations!
Our winter, as yet, has been mild, and open, and favourable

to your pursuits. Pray present my respects to your Lady, and desire her to accept of my best wishes, and all the compliments of the season, jointly with yourself. I have now squirrels in my outlet; but if the wicked boys should hear of them, they will worry them to death. There is too strong a propensity in human nature towards persecuting and destroying!

I remain, with much esteem,

Yours, &c., GIL. WHITE.

To the Rev. R. Churton.

[Endorsed by Mr. Churton, " The last from my dear Friend."]

Selborne, Jan. 26, 1793.

Dear Sir,—Had you staid only one day longer with us, you would have seen J. White* and his bride, late Miss Louisa Neave, who, having been married at Downton near Sarum by Mr. Lear, set off immediately for this place. We have good reason to be pleased with our new relation, who is sensible, intelligent, and in her carriage much of a gentlewoman. She is a nice needlewoman and also a proficient in music, and can shoulder a violin, out of which she brings a good tone, but could find no one to accompany her. Though her husband is in stature one of the sons of Anak, yet he has made choice of a little wife, who, we all agree, in her profile resembles Miss Reb. Chace, but exceeds her in her make and turn of person.

I am much obliged to you for the Latin translation of the *Caprimulgus*, which will be useful, but have lost my advocate

---

* The reader may perhaps care to learn something of the later career of Gilbert White's nephew and former pupil, " Gibraltar Jack." His first wife died in the East Indies in 1802. In 1809 he married his cousin Elizabeth, daughter of Henry White, of Fyfield, Hants, and died at her brother's house at Blakesley, Northants, on June 30th, 1821. He was buried in Maidford Church, where a monument in the chancel, erected by his widow, testifies to his talents and labours in his profession, and states that after settling at Salisbury " he continued some years" there. "He then passed some years in the East Indies, and his life was afterwards varied by many trying scenes and circumstances in different countries." He left no issue.

with the R.S.; for on my applying to Mr. Barrington, who
used to present my papers, he writes me word that he has
no longer any interest with that society, but that he will
endeavour to find a member that shall present my dissertation. This circumstance, as you may imagine, is not so
pleasant as when I had a friend who was often one of the
Council, and ready to abet my compositions.

There is, indeed, a curious coincidence of opinions between
Mr. Lewis and the Stagyrite! for which I cannot advance a
better reason than what you have mentioned yourself. Yet
can I not call that a *foolish bird* which knows the times and
the seasons, and conducts its migrations over seas and
continents with such accuracy and success, and, impelled by
all the feelings of στοργή and affection, is ready to repell
intruders, and by menaces to defend to the best of its power
its callow and helpless young!

I have told you sometimes of an old physician at South-
hampton, Dr Speed, who used to go over once every year, in
May, to the Isle of Wight, for which period the people used
to reserve their ailes. For these last two winters my coughs
have been kept till your arrival, and then became so bad
that without your kind assistance I could not have continued my duty. When you left me I had some dread
about the ensuing Sunday; but, thanks be to God, my
infirmity ceased on the Saturday, and has not been bad
since. As soon as your letter came we turned to my peerage book, but could find no traces respecting Lord Malms-
bury; so I conclude that his creation was subsequent.
Possibly before now you may have recovered your stray
idea, that has wandered away, or lay snug in some corner of
your memory.

Mrs. J. White joins in best respects and wishes to you
and all friends.   Yours sincerely,

                                        GIL. WHITE.

Sad work in France!!

Those who reside in the neighbourhood of the beautiful common, which lies above and includes the well-known Hanger, at Selborne; and even those who have only visited it, will be interested in the attempt to inclose, and therefore to destroy it, as far as the public are concerned; which is described in the following letter from one of the sons of Benjamin White, senior. The Mr. Fisher mentioned seems to have been an attorney of the baser sort, who regarded property as potential costs.

From his nephew James White.

Fleet Street, London,

Sunday, Feb. 12th, 1793.

Dear Sir,—In conversation with Mr. Fisher the other day, he told me that the scheme of the Selborne Inclosure is at an end. Now, as you are interested in this, I hope you will excuse my communicating to you Mr. Fisher's and my own sentiments. Mr. Fisher told me that the Inclosure Scheme was one of his own and not dictated to him by Magdalen College. That it was undertaken by him under an idea that they were more interested in it than has since turned out. He says that he had formed a plan and drawn an Act of Parliament for this purpose, but that you told him, some time last Summer, that you had in your possession a copy of a Decree in the Court of Chancery, made in your grandfather's time, which decreed that the Wood and Commons and other Waste Land in the Manor belonged to the Tenants and not to the Lords of the Manor. In consequence of this information from you, he says he shall not pursue his scheme, because, if this is the fact, the benefit which the College will receive from an Inclosure

SELBORNE FROM THE COMMON ABOVE THE HANGER

[To face p. 258, Vol II.

will not be sufficient to engage them in the expence and trouble which an act of this nature always occasions. Finding this, he has not acquainted the College of his intentions, so that the Fellows and other persons interested under the College, have not ever been officially acquainted that such a Scheme was ever in agitation. I believe if the plan had been pursued, that Mr. Fisher would have been more benefited by it than the College, which must convince every one (and I am sure it does me) that he is a Man of a meddling disposition, and has the interest of his own pocket more at heart than that of his Employers. He told me that though, in consequence of your information, he has given up the scheme, yet he means, when he next goes to Selborne, to ask you for a sight of this Decree, and that if he should find it anyways different from his present ideas, he shall revive it. Now, in consequence of this intelligence from Mr. Fisher, and to prevent this business being revived, I trust you will excuse my giving you a hint. If you will take the trouble to look into the Decree, and if it is what Mr. Fisher from conversation with you, supposes it to be, there can be no objection to shew it him when he asks to see it, but if there should be any thing in it which is different from his present ideas, or would induce him to revive this business, I think that you would wish to evade shewing him any thing which may, in the event, occasion much trouble and expence both to yourself and all the other copyholders of Selborne. To convince you (if it is wanted) that Mr. Fisher is a man of a meddling disposition, he acknowledged to me that he had been urging four different Inclosures of the same nature as Selborne, and intended to bring them forward this Session of Parliament, but that not one of them would be entered into on account of something similar to the reason which prevents his Scheme at Selborne. I am glad in saying that Chalgrove is one of them, which is put off this Session, but, I fear, it will be revived the next. Mr. F. told me that, upon enquiry, he finds that the major part of the copyholders at Selborne are very poor, which alone would make him very cautious. This alone sufficiently convinces me of his Motive. I wished to have waited on you to have personally informed you of the above, but could not find time during my last short visit in Hampshire. I beg my respects to Mrs. White, and

I am, Dr Sir,

Your very obliged and affectionate Nephew,

JAMES WHITE.

The writer of the above letter subsequently entered the army, and died a captain in the 82nd Regiment, in 1796, at Port-au-Prince, S. Domingo. That the inclosure scheme would be resented and resisted by Gilbert White is certain; and he may be supposed to have had other grounds than that of the resulting expence, mentioned in the letter next printed; an inference which is confirmed by the following entry made in 1789 by him in one of the Selborne Church Registers :—

" Be it remembered that there had been from time immemorial an undisputed bridle road from the east corner of the north field, across Bushy plot and along the south end of Norton mead and the north end of Yfremead, and across the seven acres into the hollow stony lane leading to Norton farm—till about the year 1770, when Sir Simeon Stuart at the instance of farmer Young, then his tenant at Norton farm, ordered the abovesaid road to be shut up and so deprived the neighbourhood of the advantage of that way: but now, in 1787, Mr. Hammond senior,

aged 81, of Newton great farm, but till late of Little
Ward-le-ham, demanded a passage for himself and horse,
of which he and others have made use the summer thro':
nor has farmer Richard Knight, the present tenant of
Norton farm, made any objection to what has been done.
The Stuart family always used to put up the wicket gates
of this road, and keep them in repair. This account was
written by Gilbert White, an ancient inhabitant and native
of Selborne, on Sept. 22nd, 1789.

> "Witness my hand,
>
> "GIL. WHITE."

To Benjn. White, senr.

Selburne, Feb. 19th, 1793.

Dear Brother,—Mrs. J. White and I return you and
Mrs. White many thanks for your kind invitation for the
25th of this month: but should be glad to defer our visit
to the Monday following, if convenient, viz. March the
4th, when, if well and able, and with permission, we will
wait on you. I propose to stay with you, if convenient,
one Sunday; but Mrs. J. White will proceed on to town, and
so return to Mareland before I leave you. Thanks for your
hint about the enclosure, of which Mr. F. wants to make
a jobb: and to bring in a bill on the copyholders of 3 or
£400, to pay which many of us must mortgage all we
possess. . . .

My neighbours have at last agreed to stint their beechen
wood, and to come to one cord for each copy: had they
used such moderation 20 years sooner, that fuel would
never have been exhausted. The crews of French privateers
now land, and plunder, as they did in Queen Anne's wars.
We rejoice to hear that my nieces have better health.
Mrs. J. White joins in respects.

> Yr loving brother,
>
> GIL. WHITE.

The shadows of life were now getting very long with the old Philosopher at Selborne. On March 4th, 1793, he left home for the last time on the proposed visit to his brother Benjamin at Mareland, Bentley, near Farnham. The thoughts of the old man, so near the end of his life, reverted to his childhood's days, when he made this entry in the *Naturalist's Journal*, which, as usual, travelled with him :—

"March 10. The sweet peal of bells at Farnham, heard up the vale of a still evening, is a pleasing circumstance belonging to this situation, not only as occasioning agreeable associations in the mind, and remembrances of the days of my youth, when I once resided in the town * : but also by bringing to one's recollection many beautiful passages from the poets respecting this tuneable and manly amusement, for which this island is so remarkable. Of these none are more distinguished, and masterly than the following :—

> 'Let the village bells as often wont,
> Come swelling on the breeze, and to the sun
> Half set, ring merrily their evening round.
>
> *   *   *   *   *
>
> It is enough for me to hear the sound
> Of the remote, exhilarating peal,
> Now dying all away, now faintly heard,
> And now with loud, and musical relapse
> In mellow changes pouring on the ear.'"
>
> "'The Village Curate.'"

"There is a glade cut thro' the covert of the Holt opposite these windows, up to the great Lodge. To this opening a herd of deer often resorts and contributes to enliven and diversify the prospect, in itself beautiful and engaging.

"Mar. 14. Took a walk in the Holt up to the lodge;

---

* If by "the town" the writer meant "*a* town," the supposition that he was at school at Farnham, mentioned *supra* vol. i. p. 29, is without any foundation.

no bushes, and of course no young oaks: some Hollies, and here and there a few aged yews: no oaks of any great size. The soil wet and boggy."

"Mar. 15. On Friday last my brother and I walked up to Bentley Church, which is more than a mile from his house and on a considerable elevation of ground. From thence the prospect is good, and you see at a distance Cruxbury hill, Guild down, part of Lethe hill, Hind-head, and beyond it the top of one of the Sussex downs. There is an avenue of aged yew-trees up to the church: and the yard, which is large, abounds with brick-tombs covered with slabs of stone: of these there are ten in a row, belonging to the family of the Lutmans. The church consists of three ailes, and has a squat tower containing six bells. From the inscriptions it appears that the inhabitants live to considerable ages.

"There are hop-grounds along on the north side of the turnpike road, but none on the south towards the stream. The whole district abounds with springs.

"The largest spring on my brother's farm issues out of the bank in the meadow, just below the terrace. Somebody formerly was pleased with this fountain, and has, at no small expence bestowed a facing of Portland stone with an arch, and a pipe, thro' which the water falls into a stone bason, in a perennial stream. By means of a wooden trough this spring waters some part of the circumjacent slopes. It is not so copious as Well-head."

The next day the visitors returned to Selborne.

To Benjamin White.   Selburne, Mar. 21st, 1793.

Dear Brother,—It begins to be full time for us to return our best thanks to you, and Mrs. White, for the kind reception which we experienced at Mareland. The day on which we left you proved so wet a one, that the country was quite drenched with water, so that in our clays the farmers can neither plow, sow, nor delve. A wet March is very unfavourable to this district, and to all strong soils, so as to occasion a failure of crops. We must therefore hope that the remainder of the spring will prove more dry, and favourable.

Timothy the tortoise came forth on the 15th instant, and has appeared almost every day since. We have planted out your cauliflowers in rich ground.

My cucumbers thrive, but are not so forward as yours: my crocus's make still a gay show, so as even to attract the attention of your granson Glyd,* who, looking at them, cries, "Pretty!"

Mrs. J. White joins in respects to you and family. Dr Chandler sets off for London to-morrow. Mr. Marsham, from whom I have just heard, does not, I find, much like Arthur Young.   Your loving brother,

                                                            GIL. WHITE.

Richard White called here this morning, and looked stout and jolly.

On April 6th the *Naturalist's Journal* records the last search for torpid swallows :—

"On the 6th of last October I saw many swallows hawking for flies around the Plestor, and a row of young ones with square tails, sitting on a spar of the old ragged thatch of the empty house. This morning Dr Chandler and I caused the roof to be examined, hoping to have found some of those birds in their winter retreat: but we did not

---

* Glyd White, the only one of Benjamin White, junior's, children who attained majority, took his M.A. degree at Oriel College in due course. He subsequently became Curate-in-charge of Ewelme, near Wallingford, of which Canon Payne Smith (subsequently Dean of Canterbury) was the Rector, in whose house at Christ Church he died in 1869 from the effect of an accident, at an advanced age. Mrs. Benjamin White, junior, died at Ewelme in 1833.

meet with any success, tho' Benham searched every hole and every breach in the decayed roof.

"April 9th. Thomas Knight, a sober hind, assures us that this day on Wish-hanger Common between Hedleigh and Frinsham he saw several Bank martins playing in and out, and hanging before some nest-holes in a sand-hill, where these birds usually nestle. This incident confirms my suspicions, that this species of *Hirundo* is to be seen first of any; and gives great reason to suppose that they do not leave their wild haunts at all, but are secreted amidst the clefts, and caverns of these abrupt cliffs where they usually spend their summers. The late severe weather considered, it is not very probable that the birds should have migrated so early from a tropical region thro' all these cutting winds, and pinching frosts: but it is easy to suppose that they may, like bats and flies, have been awakened by the influence of the sun, amidst their secret *latebræ* where they have spent the uncomfortable foodless months in a torpid state, and the profoundest of slumbers. There is a large pond at Wish-hanger which induces these sand martins to frequent that district. For I have ever remarked that they haunt near great waters, either rivers or lakes.

"April 12. The nightingale was heard this harsh evening near James Knight's ponds. This bird of passage, I observe, comes as early in cold cutting springs as mild ones!

"April 29. I have seen no *Hirundo* yet myself.

"May 1. There is a bird of the blackbird kind, with white on the breast, that haunts my outlet as if it had a nest there. Is this a ring-ouzel? If it is, it must be a great curiosity; because they have not been known to breed in these parts.

"May 5. Cock redstart. House-martin appears.

"May 7–11. James Knight has observed two large field-fares in the high wood lately, haunting the same part, as if they intended to breed there. They are not wild. A nest of this sort of bird would be a great curiosity!

"M[issel] thrushes do not destroy the fruit in gardens like the other species of *turdi*, but feed on the berries of misseltoe, and in the spring on ivy berries which then begin to ripen. In the summer, when their young become fledge, they leave neighbourhoods, and retire to sheep walks, and wild commons. The magpies, which probably have young, are now very ravenous, and destroy the broods of missel-thrushes, tho' the dams are fierce birds and fight boldly in defence of their nests."

To Dr. Loveday.

[At Williamscote, near Banbury].

Selborne, May 11th, 1793.

Dear Sir,—I sit down to return you my sincerest, though tardy thanks as the editor of Doctor Townson's 'Discourse on the Evangelical History,' etc. which gave me much satisfaction; and came the more apropos, as it arrived in Passion week.   There is a discernment in all Dr Townson's writings almost peculiar to himself.

Yet must not the Biographer go without his due share of praise; he having discovered much piety and gratitude in what he has written.   I had not the happiness of knowing Dr Townson, who, I doubt not, was a most engaging man: but I will bear testimony to the truth of what is said respecting your excellent father.   The world can ill spare such valuable characters, because few such are left behind to supply their place.

When you and your Lady visit Dr Chandler and Lady, you will I hope afford us as much of your company as is consistent with your engagements over the way at the Parsonage.   D^r and Mrs. Chandler are in London: Mrs. Chandler has lost by death, within these ten weeks, two maiden sisters, and a grandmother.   With my best respects to your lady, I remain

Your obliged, and most humble servant,

GIL. WHITE.

Do you know Dr Percy, Bishop of Dromore? Among several entire strangers I have lately received a letter from him expressing his approbation of my *Natural History* in terms that I must not repeat.

Give my respects to the Rector of Middleton,* and tell him I hope he will come and see us not long hence.

Dr. Loveday appended this note :—

" N.B.—The truly ingenious and worthy writer died at Selborne on June 26th, 1793, aged 73, Senior Fellow of Oriel College."

The *Naturalist's Journal* continues—

" May 12–18. The fern owl, or churn owl returns, and chatters in the Hanger.

" Sowed in the three-light annual frame African and French marrigolds, China asters, pendulous Amaranths, Orange-gourds.

" A man brought me a large trout weighing three pounds, which he found in the waste current at the tail of Bins pond, in water so shallow that it could not get back again to the Selborne stream.

" Took the blackbird's nest a second time; it had squab young.

" Set the second Bantam hen over the saddle cupboard in the stable with eleven dark eggs.

" A solitary hen red-start in the garden.

" Timothy travels about the garden.

" Made rhubarb tarts, and a rhubarb pudding, which was very good.

" May 22. Nep. Ben. White, and wife came.

" [May] 28. The season is so cold, that no species of *Hirundines* make any advances towards building and breeding.

* Mr. Churton.

"[May] 29. Brother Benjn and Mrs. White, and Mary White, and Miss Mary Barker came.

"June 2.   Bro. Benjn and I measured my tall beech in Sparrow's hanger, which, at 5 ft. from the ground, girths 6 ft. 1 inch, and three quarters.

"[June] 7.   Mrs. Clement and children came.'

"[June] 14.   Mr. John Mulso* came.

"[June] 15.   Mr. J. Mulso left us."

The following letter written on this day (June 15th), probably the last ever written to anyone by Gilbert White, concludes his correspondence with Marsham; which, interesting as it is to all naturalists, is especially interesting to admirers of the Selborne Naturalist as showing that he retained to the very close of his life as fresh an intellect, and wrote with as keen a relish as ever upon his favourite subject :—

To R. Marsham.   Selborne, June 15th, 1793.

Dear Sir,—From my long silence you will conclude that Procrastination has been at work and perhaps not without reason.   But that is not all the cause: for I have been annoyed this spring with a bad nervous cough, and a wandering gout, that have pulled me down very much, and rendered me very languid, and indolent.

As you love trees, and to hear about trees, you will not be displeased, when you are told that your old friend the great *Oak* in the *Holt* forest is, at this very instant, under particular circumstances.   For a brother of mine, a man of Virtù, who rents Lord Stawell's beautiful seat near the Holt, called Mareland, is at this very juncture employing

---

* The Rev. John Mulso, Vicar of South Stoneham, near Southampton; the eldest son of Gilbert White's old friend John Mulso, Canon of Winchester.

a draughts-man, a French Refugee, to take two or three
views of this extraordinary tree on folio paper, with an
intent to have them engraved. Of this artist I have seen
some performances; and think him capable of doing justice
to the subject. These views my Brother proposes to have
engraved, and will probably send a set to you, who deserve
so well of all lovers of trees, as you have made them so
much your study, and have taught men so much how to
cultivate and improve them. I have told you, I believe,
before, that the great *Holt Oak* has long been known in
these parts by the name of the *grind-stone Oak*, because
an implement of that sort was in old days set up near
it, while a great fall of timber was felled in its neighbourhood.

After a mild, wet winter we have experienced a very
harsh, backward spring with nothing but N. and N.E.
winds. All the *Hirundines* except the sand-martins were
very tardy; and do not seem even yet to make any
advances towards breeding. As to the sand-martins they
were seen playing in and out of their holes in a sand-cliff
as early as April 9th. Hence I am confirmed in what I
have long suspected, that they are the most early species.
I did *not* write the letter in the 'Gentleman's Magazine'
against the torpidity of swallows: nor would it be consistent with what I have sometimes asserted, so to do. As
to your recent proof of their torpidity in Yorkshire, I long
to see it. But as much writing is sometimes irksome,
cannot you call in occasionally some young person to be
your amanuensis?

There has been no such summer as this, so cold and
so dry, I can roundly assert, since the year 1765. We have
had no rain since the last week in April, and the first two
days in May. Hence our grass is short, and our spring-corn
languishes. Our wheat, which is not easily injured in
strong ground by drought, looks well. The hop-planters

begin to be solicitous about their plantations. Here I shall presume to correct (with all due deference) an expression of the great philosopher Dr Derham. He says in his Physico-theology " that *all* cold summers are wet ": whereas he should have said *most*.

Have You seen Arthur Young's ' Example of France a warning to England ' ? it is a spirited performance. The season with us is unhealthy.

With true esteem,

I remain, Y^r obliged servant,

GIL. WHITE.

The letter ends with ominous words.

Saturday, June 15th, is the last day on which any observation occurs in the Journal, though the days of the month were filled up for the following week ending June 22nd.

On June 10th the Curate of Selborne had been well enough to officiate at the funeral of a young girl ; when, as in the case of the last entries in the Journal, there is little or no sign of illness to be seen in the handwriting of the record in the burial register.

It became necessary, however, to send for Mr. Webb, the doctor at Alton, on June 17th, and from this date he visited his patient every day ; finding him, according to Mr. Bell's account, which was apparently made on the authority of a nephew of the Naturalist, in much suffering, borne with exemplary patience, and with the consolations of religion. That there was some pain during the closing days is confirmed by the occurrence of

THE NATURALIST's JOURNAL.

Year. Place. Soil	Therm.	Barom.	Wind	Inches of Rain or Sn. Size of Hail ft.	Weather	Trees first in leaf. Plants first in flower. Mosses first appear or vegetate.	Birds and Insects first appear or disappear. Fungi first appear or disappear.	Miscellaneous Observations with Memorandums
Selborne								
Sunday. June 9.	8 · 12 · 4 · 8		S.W. 29· 7-10·		swet clouds. fine even. cool.		early orange-lilies blow. few chafers.	
Monday. 10.	8 · 12 · 4 · 8	61.	S.W. 29· 7-10½· 16		showers. cool. sleaps: cold aur.		Cut five cucumbers.	
Tuesday. 11.	8 · 12 · 4 · 8	61.	S.W. 29· 6-10½·		grey. suñ. low'ring.		A gar. brought me a large plate of straw-berries, which were codde, & not near ripe.	
Wednes. 12.	8 · 12 · 4 · 8	62.	N. 29· 7-10·		bright. sun. golden even.		cut eight cucumbers. Mess Clement & children left us. Many swifts.	
Thursd. 13.	8 · 12 · 4 · 8	61.	N.E. 29· 8-10½·		showers– suñ. cold wind.		Cut ten cucumbers. Provence roses blow against a wall. Damsel-cherries very fine. Stocks still in full beauty.	ten weeks
Friday. 14.	8 · 12 · 4 · 8	48.	N.E. 29· 8-10·		cold wind. dark. gleams.		cut four cucumbers. M.r John Mulso came.	
Saturday. 15.	8 · 12 · 4 · 8	60.	N.E. 29· 8-10·		suñ. low'ring.		Men wash their sheep. M.r J. Mulso left us.	

" anodyne draughts " in Mr. Webb's account, which happens to be in the writer's possession. Some time previously the bed had been moved into the old family parlour on the first floor, at the back of the house; and the last scene the Naturalist's dying eyes must have looked upon was his garden and fields, with the trees, many of which he had himself planted, and beyond them the beautiful beech-crowned Hanger.

On June 25th a visit both in the morning and evening was paid.

On the 26th an express messenger was sent to Salisbury for Dr. John White. He posted to Selborne at once, but can hardly have found his uncle alive; since on the latter day the White family lost its amiable head; Selborne a highly respected neighbour; and the world a singularly observant and original naturalist.

What is the happy life? It is a true, if trite, saying that few men attain their ideal of a career in life; or, having attained it, realise that it is the ideal career. But the man who lay dead at Selborne, fascinated from boyhood by the study of Nature, had longed for life and leisure in his wild, woodland, native country—not from any merely indolent wish to shirk the responsibilities of life, to cope with which he was by character and attainments amply equipped—of him it may be truly said that he had realised his ideal, and as much as any man had lived a happy life.

CHAPTER XII.

WITH characteristic good sense Gilbert White had objected, in Letter V. of his 'Antiquities of Selborne,' to "the improper custom of burying within the body of the church," though so many of his kindred were laid in the chancel at Selborne. His will directed as follows :—

"And lastly to close all I do desire that I may be buried in the church yard belonging to the parish Church of Selborne aforesaid in as plain and private a way as possible without any pall bearers or parade and that six honest day labouring men respect being had to such as have bred up large families may bear me to my grave to whom I appoint the sum of ten shillings each for their trouble."

Mr. Taylor paid his curate the compliment of a visit to Selborne to bury him, the funeral taking place on July 1st.

By his will Gilbert White bequeathed £100 to Oriel College "as a small acknowledgement for the many favours I have received from that Society for near a half a century past." Legacies were left to his nephews and nieces : to his brother Benjamin he bequeathed his (copyhold) "dwelling-house and appurtenances known formerly by the name of Wakes." His sister-in-law, who had for some years resided with him, received £200, the household goods and furniture, and a small annuity; and to his friend the Rev. Ralph Churton was left a valuable copy of Bishop Tanner's 'Notitia Monastica.' The rest of his library was divided between his nephews, John White, surgeon, and the Rev. C. H. White, son of the Rev. Henry White. "My old servant Thomas Hoar" was not forgotten.

When the Provost of Oriel heard of the death he wrote to Benjamin White, senior, "Your son Edmund was so kind as to inform us of your and our great loss. Your brother's death was, I will assure you, most sincerely regretted by the College and will long continue to be so." A little later he wrote again to acknowledge the receipt of the legacy to Oriel College, "We shall take care that your Uncle's kind remembrance of us shall not be forgotten. His memory will ever be respected by his Oxford friends, and dear to those of his own College; at least I am sure it will ever be so to your very faithful and obedient servant, J. Eveleigh."

A letter, signed "A Southern Faunist," dated 11th July, 1793, appeared in the 'Gentleman's Magazine':—

"A sigh escapes me on the demise of that most excellent man, accurate historian, diligent naturalist, and elegant

writer, the Rev. Gilbert White. I hope a monument will
be erected to his memory in the church he has so pleasingly
described, in which I conclude he is interred."

The members of his family placed the now well-
known tablet to their relative's memory outside the
north wall of the chancel, in proximity to the grave,
which was marked by a headstone, bearing the
initials " G. W." and date of death, " 26 June,
1793."

" In the fifth grave from this wall are interred the Remains of
the Revd. Gilbert White, M.A.,
Fifty years Fellow of Oriel College in Oxford,
and Historian of this his native parish.
He was the eldest son of John White Esquire Barrister-at-Law
and Anne his wife, only child of
Thomas Holt, Rector of Streatham in Surrey.
Which said John White was the only son of Gilbert White
Formerly Vicar of this Parish.
He was kind and Beneficent to His Relations
Benevolent to the Poor
And deservedly respected by all his Friends and Neighbours.
He was born July 18th, 1720, O.S.
And died June 26th 1793.
Nec bono quicquam mali evenire potest,
Nec vivo, nec mortuo."

In 1810 this monument, together with one to
Benjamin White the elder, was removed for the sake
of better preservation into the chancel. Unfortunately, presumably because the chancel wall next
the graves was fully occupied with tablets to the
White and Etty families, they were placed on the
side furthest from the graves; so that the indication
given in them of the position of the latter is now

"Then let wise Nature work her will,
    And o'er my clay her darnel grow;
Come only, when the days are still,
    And at my headstone whisper low,
    And tell me if the woodbines blow."

[To face page 274, Vol. II.

misleading, and has constantly puzzled those admirers of the Naturalist's graceful writings, who, when on a pilgrimage to Selborne, never omit to seek out the grave of Gilbert White.

Among the many distinguished men who have made the Selborne pilgrimage was the late James Russell Lowell, who happened to be staying in the neighbourhood with Lord Selborne in 1880, when he wrote the following verses for his host :—

> " To visit Selborne had been sweet
>   No matter what the rest might be ;
> But some good genius led my feet
>   Thither in such fit company,
> As trebled all its charms for' me.
>
> " With them to seek his headstone grey,
>   The lover true of birds and trees,
> Added strange sunshine to the day.
>   My eye a scene familiar sees,
> And Home ! is whispered by the breeze.
>
> " My English blood its right reclaims ;
>   In vain the sea its barrier rears ;.
> Our pride is fed by England's fame,
>   Ours is her glorious length of years ;
> Ours, too, her triumphs and her tears."

The writer of this biography is doubly disqualified from composing what is called an "appreciation" of Gilbert White's life and work ; because he happens to be a relative, and because he is no naturalist. Yet no person of ordinary intelligence can peruse 'The Natural History and Antiquities of Selborne' without forming some estimate of its author's abilities and position in the literature of natural history.

The question is frequently asked, Why has this book, alone among books of its class on natural history, lived? Why does it even still appeal to so many men of so differing tastes and positions in life? Why is it constantly republished? One reason may be that it is written in a style which may be called a model of clear unaffected English. Its author did not, as a boy and young man, enjoy the benefit of any instruction in facts of science; instruction which would in our time be dignified with the name of a scientific education; but his reasoning faculties were strengthened and improved by the study of the classical languages, which not only introduced him to the noblest literature, but also taught him to be logical and careful in thought, and accurate in statement.

To criticise the book in the light of modern knowledge would be absurd. Yet the thought has often occurred to the present writer, when reading it, that Gilbert White was no unworthy forerunner of that greatest of naturalists of our time, who is said to have changed the thoughts of men. Perhaps when the former wrote "the two great motives which regulate the proceedings of the brute creation are love and hunger"* we may not find in this even the germ of Darwin's great theory of sexual and natural selection; but "protective mimicry" is clearly in-

---

* *Vide* 'The Natural History of Selborne,' Letter XI. to Barrington.

dicated in the history of the stone-curlew,* whose young "are withdrawn to some flinty field by the dam, where they skulk among the stones which are their best security; for their feathers are so exactly of the colour of our grey-spotted flints." And in the well-known remarks upon the work and use of earthworms † the philosopher of Selborne did something more than "throw out hints, in order to set the inquisitive and discerning" philosopher of Downe to work at a paper, which much resembles a "good monography of worms," and was read by him before the Geological Society, just sixty years after the former had written his letter to Daines Barrington.

The popularity of the book undoubtedly owes much to its subject. Outdoor life is ever sought after by Englishmen, and perhaps on the principle that

"There's not a joy the world can give
Like that it takes away,"

this book of outdoor life loses nothing from the fact that we are becoming more and more a nation of townsmen. As an illustration of this feeling, a well-known London solicitor, the late Mr. Edward Tylee, told the present writer that, when he left a country home as a youth and came to work in a London office, the only alleviation of the great change from his country life was reading 'The Natural History of Selborne' at breakfast-time.

* *Op. cit.*, Letter XVI. to Pennant.
† *Op. cit.*, Letter XXXV. to Barrington.

But the great glory of the book is that it has stimulated so many young people to make a profitable use of their powers of observation, and, by studying the natural objects around them, to live happier and fuller lives. As a typical instance of this there are two men in the county of Kent, whom circumstances have placed as tradesmen in a small town and village respectively. Each of them has ennobled a life of commerce, and enriched scientific knowledge by his devotion, the one to the geology of the Tertiary strata near Sheerness and to marine zoology; and the other to an exhaustive study of the very rude flint implements of early man on the plateau of the North Downs, to which he was the first to draw attention. These gentlemen have stated to the present writer that when young men their thoughts were led to observe matters of interest in their locality by 'The Natural History of Selborne'; thus fulfilling, as doubtless very many others have done, the aspiration of Gilbert White, expressed in the Preface to his book, that he might have "induced his readers to pay a more ready attention to the wonders of Creation."

How far Gilbert White deserves to be called a great naturalist may be the subject of argument, of doubt; but there can be none that to have taught his countrymen how and what to observe is to have done a very great thing.

It is greatly to be regretted that, though Thomas White urged his brother to sit for his picture, no portrait or sketch of any kind was ever made of him. It is known, however, that he was five feet three inches in stature, and slender in person. He is said to have possessed a very upright carriage and a presence not without dignity. If he resembled his brothers, his features were regular, his complexion fair, and his eyes brown. The expression of his countenance was intelligent, kindly, and vivacious. In default of any portrait we can only

> "look
> Not on the picture, but the book,"

in which, indeed, its writer's thoughts and character are clearly mirrored.

The youngest of those who knew Gilbert White has long been dead, and but little is on record of his habits. But it is known that he was in every sense of the word a gentleman; kind and courteous in manner, and liberal in pocket to his poorer neighbours; and he is still spoken of in Selborne as having been especially devoted in his attention to his sick parishioners. Perhaps his most prevailing characteristic was caution—not, however, unmixed with candour—which, indeed, is amply apparent in some of the letters now printed for the first time; and shrewd practical common sense, which does not always accompany studious habits, was not wanting in his case.

These pages will have been written to little
purpose if they have not made abundantly evident
the affection which Gilbert White bore to his native
village. Indeed, the same may be said of his
brothers, though circumstances sent them early afield.
Thomas, as soon as he was enabled, by the inheritance of a considerable property, to retire from
business in 1777, constantly visited at Selborne;
where, as has been mentioned, he acquired property. His copy of his brother's book, which has
descended to his great-grandson, the writer of this
memoir, has printed across its cover the following
quotation from Guarini's 'Il Pastor Fido,' act ii.
sc. 5 :—

"CARE SELVE BEATE,
E VOI SOLINGHI E TACITURNI ORRORI,
DI RIPOSO, E DI PACE ALBERGHI VERI.
O QUANTO VOLONTIERI
A RIVEDERVI I TORNO."

"Dear happy groves, the true abode of solitary and silent
awe, of repose and peace. O how willingly would I return
to see you again!"

Benjamin, as has been seen, came to reside in
the neighbourhood of his old home; and Henry,
who created a duplicate of some of the Selborne
amenities for himself at Fyfield, was often at Selborne, his native place.

Nor has Selborne forgotten the man who made
her name a household word.

Many memorials to her Historian are there to

be seen.   About fifty-eight years ago a village school
was built in his memory, and recently water from
the perennial spring, "Well Head," has been laid
on to the village street as a memorial of the
centenary of his death—a work in which he would
surely have taken great interest.   At the centenary
meeting in 1893 the question of erecting a statue
of the Naturalist was debated, but the idea was
negatived.   Perhaps it was thought,

"Si monumentum requiris, circumspice."

The name of his native village has indeed given
a title to one who was worthy of honour; but while
the name of Selborne remains there is no need of
a statue to cause it to be for ever associated with
a good man and a distinguished Naturalist—with
GILBERT WHITE.

The brothers of Gilbert White living at his death
did not very long survive him.   Benjamin died at
his residence, Mareland, in March of the following
year, 1794.   He was buried near his brother at
Selborne.   Thomas died in February, 1797, and was
buried at Harlow, Essex, near which place he owned
a manor and estate.*

Thomas Hoar, the Naturalist's faithful old servant,
died in April, 1797, at the ripe old age of 83.
Another who may be called a member of the house-

* The grave marked "T. W." near Gilbert White's is that of a son of his
nephew and niece, Mr. and Mrs. Benjamin White, who died in boyhood in
1795.

hold, Timothy, the tortoise, is said to have died in the spring of the year following his owner's death. His shell is now to be seen in the British Museum (Natural History), Cromwell Road.

The fortunes of the publishing firm were not so successful under Benjamin and John White as in their father's time. The former died in 1821, and the latter, who lived at Selborne, in a house (now long pulled down) which he built in the grounds of "The Wakes," had serious losses from the defalcations of a manager, the business being ultimately sold. The last of the family to reside at Selborne was Mary, the only daughter of Benjamin White, senior, who remained unmarried. She occupied "The Wakes" for many years; latterly with a niece, Georgiana, daughter of her brother, John White; dying, in 1839, while on a visit to her nephew, the Rev. Herbert White, Vicar of Warborough, Oxfordshire, whence she was brought to Selborne for burial.

The house then stood empty for some time in a neglected state. It was put up to auction in July, 1840, by the celebrated George Robins, who described its attractions and association with Gilbert White in his usual florid style. It was ultimately purchased in 1844 by the late Professor Bell, who carefully and judiciously preserved every memorial of Gilbert White. He found it necessary to add one more room to the west end of the house, which was done in complete harmony with the existing part. Since his death in 1880 the property has more than once

Particulars

LOT 1.

THE SELBORNE ESTATE

Is now associated with the History of England, and the interest excited by this Sale will necessarily be materially increased in public estimation by a delightful reminiscence, that it was within the hallowed walls of the Property under consideration that

THE REV. GILBERT WHITE,

THE CELEBRATED NATURAL HISTORIAN,

Achieved all his fame, thus giving au *eclat* to

THE VILLAGE OF SELBORNE,

That hath, for many a long year, placed it high in favourable opinion. Miss Mitford, in her delightful "Village," must have had Selborne throughout in her mind's eye, and the composer of the feeble attempt that is to follow, is not insensible to his inadequacy to do it justice. It is not intended to go at length into the history of a Village which owes so much of its fair fame to the talented Gentleman under review. It would fill a folio volume only to extract the varied panegyrics so justly paid to his great literary research. His first edition of " The Natural History and

ANTIQUITIES OF SELBORNE"

Is a work destined, from the great simplicity of its style, the calm benevolence of its spirit, and the close observation evinced in every page, to be the most popular of any publication that has followed or preceded it. There is a worthy divine who has located in our delightful Village for many a long year (the Rev. W. Cobbold). The Purchaser of this famed Property will find in him an invaluable Neighbour.

The Property includes, first, the Abode of the gifted person already alluded to. The Residence, in itself, has very little pretensions to the honour thus awarded to it; but the situation is so indescribably beautiful, that it will not be doubted a successor will soon create anew, by tact and judgment, all its acquirements in the olden times. It is in the centre of the Village, and most unpretending in its outward character. It is to the scene of loveliness in the rear that especial attention is directed—there is a delightful little Park, beautifully studded with Timber, grouped in the most picturesque form. A Hanging Wood, of superlative beauty, is in direct communication with the Park. The ascent is of fearful height, but there are easy paths to enable even a timid adventurer to ascend without difficulty ; and, when the task is accomplished, the splendid Panorama from above will compensate tenfold for the labour that has been induced.

> " See Selborne spreads her boldest beauties round,
> The varied valley, and the mountain ground.
> Wildly majestic."

THE HANGING WOODS,

Which approximate upon the Park, form from this summit one of the most incomparable views that England, all over, can produce—the Panorama is one of extraordinary beauty, variety, and extent, overlooking Farnham and Guildford to the Hogsback.

The Purchaser of this Lot will be entitled to all the Fixtures belonging to the Vendors (except the Stone Pedestal in the garden, and the Book-case in the Dining-room of the ancient family abode).

The Purchaser of this Lot will also be entitled to the customary allowance of Cord Wood and Faggots, from Selborne Hill, and which has usually consisted of four Cords and a half of Wood, and four hundred and a half of Faggots, annually; and also to such customary right as the Vendors possess to turn out Sheep and Beasts on Selborne Common, in respect of Twelve Common Rights.

EXTRACT FROM THE "PARTICULARS OF SALE OF THE SELBORNE ESTATE,"
JULY 25, 1840

[To face p. 282, Vol. II.

changed hands, the house has been greatly enlarged, and the old part of it materially altered.

It is so long since 'The Natural History and Antiquities of Selborne' appeared, that the book has now a history of its own, and therefore a short account of the earlier editions may be given.

The only complete edition published during the author's lifetime was the first, of 1789. This was issued at the price of one guinea, in boards. In 1792 a curiously compressed translation was published in Berlin, a fact which sufficiently attests the early success of the book.

In 1795 B. and J. White issued 'A NATURALIST'S CALENDAR with Observations in various branches of Natural History, extracted from the papers of the late Rev. GILBERT WHITE, M.A., of Selborne, Hampshire, Senior Fellow of Oriel College, Oxford. Never before published.' This consisted of extracts from the unpublished manuscript and journals of the Naturalist, edited by Dr. Aikin, 8vo; and also in large paper, 4to, probably for binding with the original book.

In 1802 J. White issued 'The Natural History of Selborne' (omitting 'The Antiquities') together with the 1795 publication, in two volumes, 8vo.

In 1813 a very handsome edition of the original book was issued by White, Cochrane, and Co., in 4to, at the price of two guineas and a half. This is printed on much better paper than the first edition.

It contains all the original plates, and, in addition, one of a view of "The Wakes" and of a picture placed over the Communion table in Selborne Church by Benjamin White, senior. Some copies of this edition were issued in large paper, with the plate of the picture mentioned coloured, at five guineas.

In 1822 and 1825 what were practically reprints of the 1802 edition were published. Since the latter date the editions have been very numerous, and need not be here referred to.

A few short extracts from some of the earlier notices of Gilbert White's book may be of interest.

From a memoir of Dr. Aikin, by L. Aikin, 1823, vol. ii. p. 194—

"This picture is equally natural with the former, and has the additional merit of furnishing new images to the fancy. It was from such a mature and deliberate study of Nature, that Mr. White of Selborne derived that store of curious observation which he has presented in the most entertaining miscellany of Natural History that was ever composed."

From the 'St. James' Chronicle,' October 6th, 1827—

"Mr. Phillips' late publication entitled 'Pomarium Britannicum' will be found perhaps the most pleasant book extant on this interesting subject, partaking in a great degree of that charm which has conferred lasting popularity on Mr. White's 'Natural History of Selborne.' . . . ."

From a memoir, by Archdeacon Churton, of Dr. Chandler, prefixed to a republication of the latter's 'Travels in Asia Minor'—

" . . . [Gilbert White's] History of his native parish of Selborne, Hants, which having since been published in more than one edition, and finding an encomiast in every reader, needs not here be commended."

From the 'Retrospective Review,' vol. xiv. p. 3—

"It has been well remarked that to De Foe's 'Robinson Crusoe' we owe more gratitude for that persevering energy of adventurous spirit which has and for ever we trust will animate our navy, than to any other cause whatever. We believe the fact, and to White's 'Natural History of Selborne,' by parity of reasoning, we feel inclined to assign the merit of that increasing attachment to the study of natural history which, since his day, has been making such rapid strides. But as De Foe was indebted to another for his invaluable fiction, so to the work before us ['The Philosophical Correspondence of Ray and Willughby'] we may ascribe the origin of Mr. White's more popular performance. True it is, that the lively and natural style of the latter must ever prove a formidable rival to its venerable precursor. . . ."

From the 'Quarterly Review,' January, 1828—

"White's delightful work is no longer shut up in a quarto. It is most pleasing to witness the exertions made by eminent writers of our time to produce food for the juvenile mind. Shall we be pardoned for observing that 'The Natural History of Selborne' ought to have a place among the household books of every English family?"

From the 'Quarterly Review,' April, 1829, review of 'The Journal of a Naturalist'*—

"We believe very few books on the subject of Natural History have met with such unqualified praise from those

---

* Published anonymously by Murray, 8vo, 1829, but known to be by J. L. Knapp, of Alveston, in Gloucestershire.

to whom the contemplation of the various objects of nature
can afford rational amusement, as 'The Natural History of
Selborne,' by the Rev. Gilbert White. The author of the
little volume, with the modest title, now before us, admits
that, in the collection of his own materials, he had this
interesting book in his eye ; that the perusal of it early
impressed on his mind an ardent love for all the ways and
economy of nature ; and that he was thereby led to the
constant observance of the various rural objects with which
he was surrounded."

So far the earlier notices of Gilbert White's book
established and confirmed his fame. The cult of
Gilbert White and Selborne may, however, be said
to have commenced in earnest with the publication of an article in 'The New Monthly Magazine,'
1830, part ii. pp. 564–570, by an anonymous
writer. This contains a vivid description of the
village of Selborne and the Hanger, from whence,
"seated in an arbour which has been formed about
half-way up" (what Gilbert White called the *new*
Hermitage), the traveller contemplated Gilbert
White's house and grounds.

The chapter in which Mr. Jesse* quotes this
interesting notice of Gilbert White concludes with
some lines which were addressed to the wife of a
nephew of the Naturalist by her father, Mr. G.
Tahourdin, upon what he terms

"The shades of old Selborne so lonely and sweet."

* *Vide* 'Gleanings in Natural History,' 2nd series, 1834.

The writer cannot more fitly bring this memoir of his kinsman to a conclusion than by quoting, by permission, the following pleasing verses, embodying as they do a present-day appreciation of the place and its Historian, which appeared in 'The Speaker' of June 17th, 1893, entitled—

"IN THE COUNTRY OF GILBERT WHITE.

" Ghosts of great men in London town
    Confuse the brains of such as dream,
But here betwixt this hanging down
And this great moorland, waste and brown,
    One only reigns supreme.

" In Wolmer Forest, old and wide,
    Along each sandy pine-girt glade
And lonesome heather-bordered ride,
A gentle presence haunts your side,
    A gracious reverend shade.

" And as you pass by Blackmoor grim,
    And stand at gaze on Temple height,
Methinks the fancy grows less dim :
Methinks you really talk with him
    Who once was Gilbert White.

" For yonder lies his own true love,
    His little Selborne, dreaming still,
The shapely ' Hanger ' towers above,
Girt with its beautiful beech grove,
    Like some old Grecian hill.

" And there th' abrupt and comely ' Nore '
    Guards that wild world of bloom and bird
Where his clear patient sense of yore
Conned sights and sounds, which ne'er before
    Sweet poets saw or heard.

" And here, hard by, the nightingale
   For the first time in springtide sang,
While Gilbert listened ; here the pale
First blackthorn flowered, while down the gale
   The cuckoo's mockeries rang !

" And there rathe swallows would appear,
   To whirl on high their last gavotte ;
And there the last of the great deer
Fell on a winter midnight clear,
   'Neath a ' night-hunter's ' shot.

" We know it all !   Familiar, too,
   Seems this quaint hamlet 'neath the steeps—
House, ' Pleystor,' church, and churchyard yew,
And the plain headstone, hid from view,
   Where their historian sleeps.

" 'Twas just a century gone by
   They laid the simple cleric here :
Th' old world was in her agony,
And ' Nature ! Reason ! ' was the cry
   In that historic year.

" But O ! another nature 'twas
   That ruled him with her magic touch,
A mistress of delightful laws,
Whom still we learn to love because
   We love her servant much ! "—Victor Plarr.

FINIS.

THE GRINDSTONE OAK (*see p.* 269)

INDEX

294 INDEX
Mouse, the harvest, I. 155, 243 *note*.
Mulso, Edward, I. 38.
Mulso, Miss Hester (Mrs. Chapone),
I. 54, 218; her marriage, 38, 118;
on the 'Invitation to Selborne,' 65.
Mulso, Miss Hester, II. 125; letter
from Timothy the Tortoise, 126–9.
Mulso, Rev. John, I. 26; his friendship with Gilbert White, 37;
letters to him, 38 *et seq.*; on his
coach-sickness, 46; M.A. degree,
47; expedition to Cowper's Hill,
56; visit to Devonshire, 60; on
the 'Invitation to Selborne,' 64,
97; the zigzag path, 71; encomium
on John White, 79; the loan of
a horse, 80; report of Provost
Hodge's death, 83; his marriage,
88; on the Provost's election at
Oriel, 91; the death of Gilbert
White's father, 98; on his retention of his Fellowship, 99; incumbent of Thornhill, 114; on the
living of Selborne, 126–8, 140;
Tortworth, 147; on his matrimonial intentions, 150; appointed
vicar and rector of Witney, 151;
on the living of Cromhall, 154;
visit to Selborne, 172, 181, 287;
canon of Winchester and rector
of Meonstoke, 204; on Grimm's
drawings, 324; presented to the
rectory of Easton, 325; on Gilbert
White's attack of illness, II. 9;
the view of the Hermitage, 21, 22;
antiquities of Selborne, 33, 36,
83; the living of Ufton Nervett,
36, 37; the death of Mrs. Snooke,
42; of John White, 66; of his
uncle, 71; the drawing of the
Temple, 74; opinion of Gilbert
White's verses, 101, 112, 146; the
death of Mr. Yalden, 143; the
new library at Oriel College, 165;
on the publication of 'The Natural
History of Selborne,' 170, 184;
receives a copy, 189; criticism of
it, 190; on the death of Henry
White, 192; of his wife, 226; his
death, 227.
Mulso, Rev. John, vicar of South
Stoneham, II. 268 *note*.
Mulso, Thomas, of Twywell, I. 37.
Mulso, Thomas, his marriage, I. 38,
118; at Selborne, 150.
Musgrave, Mr. Chardin, I. 58, 74;
elected Provost of Oriel, 91; on
Gilbert White's retention of his
Fellowship, 99, 101; his death,
157.

Musgrave, Sir Christopher, I. 91.
Musgrave, Sir Philip, I. 74, 91.
Myrmeleon (lion pismier), I. 179.

N

'Naturalist's Calendar,' II. 283.
Naturalist's Journal, entries in, I.
156, 169, 172, 175, 204, 209, 216,
231, 235, 242, 252, 278, 282, 291,
314, 324; II. 1, 9, 17, 23, 27, 30,
42, 50, 58, 65, 67, 71, 73, 75, 84,
88, 108, 109, 110, 124, 133, 139,
143, 146, 155, 162, 164, 166, 169,
171, 183, 187, 197, 207, 218, 223
note, 226, 234, 262, 264, 267.
'Naturalist's Summer Evening Walk,'
poem, II. 112, 195, 236 *note*.
'Nature Notes,' extract from, I. 4
note.
Neave, Miss Louisa, her marriage,
II. 256.
Newbury, Sampson, on the birds of
Devonshire, I. 252–4.
Newton Valence, I. 19.
Newton, Prof., on the fourth edition
of Pennant's 'British Zoology,' I.
309.
Nightingales, migration of, II. 242.
North, Bishop Brownlow, his paper
of questions to each incumbent,
II. 174.
Nuthatch, habit of, II. 67.

O

Oak, The Grindstone, II. 241, 269.
Oak, Holt, size, II. 221, 223; views,
269.
Oaks, size of, II. 228, 236.
Œstrus bovis, malady occasioned by,
II. 206, 224; *curvicauda*, I. 264,
274.
Oriel College, Oxford, I. 33; election
of Fellows, 36, 89; disputes at,
228; statutes, II. 43.
Oxford University, contest for the
Chancellorship, I. 127 *note*; claim
of the city at the coronation of
Charles II., I. 10.
Oxfordshire, Whites of, I. 7.

P

Pallas, Dr., his 'Travels,' II. 155
note.
Partridges, Spanish and Barbary,
difference between, I. 199.
Pembridge, Sir Fulke, I. 3.
Pennant, Thomas, letters from Gilbert White, I. 151 *et seq.*; various
editions of his 'British Zoology,'

PLYMOUTH
WILLIAM BRENDON AND SON, PRINTERS.

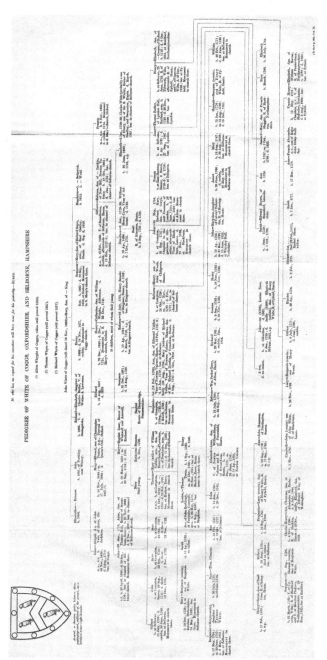

PEDIGREE OF WHITE OF COGGS, OXFORDSHIRE, AND SELBORNE, HAMPSHIRE

The material originally positioned here is too large for reproduction in this reissue. A PDF can be downloaded from the web address given on page iv of this book, by clicking on 'Resources Available'.

Printed in the United States
By Bookmasters